NASA
Range
Rats

The True Beginnings

Edward Ehrenspeck

with D. A. Hepker

Editing, Layout, and Cover Design by Sheila R. Boyd, EdD, sheilaraeboyd@gmail.com.

Dedication

To the early space workers and astronauts
who sacrificed immensely
to achieve the benefits and successes
of today's space exploration.
Never forget them or their sacrifices.

Table of Contents

Chapter 1

It is not uncommon today to hear or read about multiple rockets being launched into space each week. Of course, the term, "routine," should never be used to describe rocket launches and space exploration as it remains an extreme risk to the brave individuals traveling away from earth to explore and advance our knowledge of the unknown. In 1957, the launch of one satellite shook up the world; yet today, rockets are being launched by NASA or private companies capable of delivering up to 100 or more satellites into orbit at a time.

The rapid advancement in technology amazingly now has astronauts and civilians riding aboard these spacecrafts pursuing improvements in mankind. It certainly was not always this way, and these remarkable accomplishments in space exploration were made possible by the unknown and regrettably forgotten dedicated men and women who selflessly lived in less than luxurious surroundings and worked in conditions which today would seem primitive. These devoted workers loved their country and were dedicated to advancing the United States in the development of the space race between the U.S. and Russia and in future space exploration.

The beginnings of the U.S. space program are inexorably tied to the grim origin of the V-2 rocket. Developed in Germany as a desperate attempt to turn the tides of the war, the V-2 missile was a paradigm shift in rocketry and included the most advanced, programmable guidance system ever developed. The design was the culmination of a lifelong obsession with spaceflight by a young engineer named Wernher von Braun.

During the 1930s the German military began seeking out new weapons. Walter Dornberger, an artillery officer, was ordered to investigate the feasibility of rockets. Dornberger contacted a German rocket society and soon came in contact with von Braun.

Von Braun was born into a wealthy family and enjoyed the kind of childhood which allowed for "flights of fancy" to become true passions. On his 13th birthday, von Braun was gifted a telescope by his mother which catalyzed his love affair with the stars. He excelled in physics and math, often at the expense of his other studies. After graduating high school, von Braun joined an amateur rocket society which caught the attention of the German Army in 1932.

Dornberger recruited von Braun to help in developing liquid-fueled rockets for the military in 1932. He was just 22 years old when, in 1934, he began work with the military that would eventually lead to the creation of the V-2 missile or Vergeltungswaffe 2 (Vengeance Weapon). Produced by forced labor at Mittelwerk factory, the rocket was estimated to have killed over 2,500 during its use while over 20,000 individuals perished in its production.

It is not fully known if von Braun was personally aware of the deplorable conditions at the factory which produced his rocket. However, he was once arrested for suggesting after many drinks one evening, that the war would not end well for Germany and that his allegiance was solely rooted in service to furthering his vision of human spaceflight.

Von Braun created small rockets until late 1934. The rockets were somewhat successful, and von Braun's team moved to a larger German facility at Peenemunde on the Baltic coast. Three years later they launched the first A4 rocket that resulted in a rocket that could hit a target the size of a city at a range of 200 miles.

Hitler, however, was not particularly enthusiastic about the rocket program. He believed the weapon was simply a more expensive artillery shell with a longer range at a higher cost. As World War II continued, Hitler finally warmed to the program and, on December 22, 1942, authorized the A4 rocket to be produced as a weapon. The first production of these rockets was completed in early 1944. At that time, the A4 was redesignated the V-2 Rocket, the "V" standing for vengeance.

Due to many allied attacks at V-2 production areas, the Germans built an underground production plant with a 900,000-square-foot production area. The plant was constructed in two parallel tunnels 500 feet apart, each a mile and a quarter long and cut completely through a mountain. The main rocket assembly line started at one end of the first tunnel, and the rockets moved along on rails. The rockets were finished and tested upon reaching the opposite end of the tunnel. They were then ready for delivery to launching sites. The second tunnel was used for bringing in units and parts for subassembly lines. There were 46 smaller tunnels cross-connecting at strategic points of the two main rocket arteries. Subassemblies were channeled through tunnels in time so that they reached the main assembly line at the time and place required. The total length of the entire tunnel web was 18 miles.

The V-2 rocket was the first ballistic missile to achieve suborbital spaceflight. Over 3,000 V-2s were launched against England, France, and other Allied targets in World War II, resulting in the death of an estimated 7,250 military personnel and civilians with over 6,000 injured. An estimated 60,000 prisoners worked in factories building theV-2 rocket of which 20,000 prisoners died constructing the V-2 rocket: 9,000 died from exhaustion and collapsed; 350 were hanged, including 204 for sabotage; the remainder were shot or died from disease or starvation. The V-2 rocket is the only weapon system to have more deaths caused by its production than by its deployment.

The V-2 program became the single most expensive development of the Third Reich; and at the end of the war, German scientists were working on chemical and possibly biological weapons to be used with the V-2 rocket. None of these weapons were ever used. The last V-2 rocket was launched March 22, 1945, which killed a British citizen in her home.

The war with Germany ended May 8, 1945. At the end of the war, the race began between the United States and Russia to retrieve as many V-2 rockets and staff as possible. Before the Soviet Union arrived at their post-war occupation area, U.S. Army ordnance teams had relocated V-2 schematics and thousands of components out of Mittelwerk to the White Sands Missile Range in New Mexico where the first U.S. launch tests would be conducted. Three hundred trainloads of V-2s and parts were captured and shipped to the United States along with 126 of the principal designers.

The Russians also captured a number of V-2s and staff. In 1946 a group of just under 250 engineers were in the U.S.S.R. The first Soviet missile was the R-1, an exact copy of the V-2.

Von Braun, his brother, and seven others decided to surrender to the United States military to ensure they were not captured by the advancing Soviets. Von Braun and over 150 other skilled German rocket scientists were brought to the U.S. under the highly controversial "Project Paperclip," a program which pardoned Nazi scientists in exchange for their work in the United States. Most of the German team were sent home after that project was completed, but some remained to do research until 1951. Unbeknown to the German staff, work immediately began on larger rockets, the R-2 and R-5, based on the extension of the V-2 technology.

Both von Braun and Dornberger were now in American hands. For 15 years after World War II, von Braun worked with the United States Army in the development of ballistic missiles. Certainly, the greatest impact on modern U.S. rocketry occurred when the bulk of German rocket scientists surrendered to U.S. Forces. In spite of the history of both the V-2 and its creator, the pioneering steps taken with the technology and science developed for war continued to provide inspiration and awe to generations of individuals who looked to the stars and felt the pull of exploration.

Von Braun was one of those individuals and lived to see his dream for humankind's first steps into the cosmos become reality. He saw his work align with his vision for a spacefaring civilization. During the first episode of *Tomorrowland* in 1955, Walt Disney and von Braun brought his vision to life for the American public for the first time. By 1960, von Braun was working with the newly established National Aeronautics and Space Administration (NASA), refining his V-2 technology into the massive Saturn V rocket that would put Americans on the moon in 1969.

By March 1946 the first captured V-2 was "static fired" at White Sands, and in April 1946 the first one was launched. In April 1946 to 1951, the Army generated an increasing expertise in rocket technology and launched 67 V-2s from White Sands, establishing high-altitude and velocity records that reached the very edge of space. From these experiments emerged the first generation of American-built rockets such as the "Redstone" rocket with which I became involved in 1956 when I was first employed at the Cape Canaveral Test Range. Launch Complex 33 at White Sands tested and launched the very first generation of rockets that enabled Americans to probe into deep space.

Since 1985 Launch Complex 33 has been a national historical landmark. Two years earlier in 1983, the complex was named a State of New Mexico Historical Monument. A plaque at the site reads "Where It All Began."

The success of 67 V-2 rocket launches at White Sands was very rewarding until May 30, 1947.

> *The El Paso Times* wrote, "El Paso and Juarez rocked Thursday night when a runaway German V-2 rocket fired from the White Sands proving ground in New Mexico, crashed, and exploded on top of a rocky knoll three and a half miles south of the Walrus business district. The giant missile burst in a desolate area of jagged hills, gullies, and boondocks. No one was injured. There was an error in judgment in not shutting off the rocket motors or to the rocket destruct as soon as it was determined the rocket had swerved off course. The violent blast shook virtually every building in both El Paso and Juarez. The terrific impact of the rocket scooped out a perfectly rounded area about 50' in diameter and 24' deep. This site was a half mile from the Buena Vista airport, where 13 planes were shaken by the blast, and a mile and a half from an oil plant, where it was first believed the oil plant had exploded. Flames shot into the air like a mushroom. It looked just like a haystack on

fire. One report stated, 'I saw the rocket go right over the house. It looked like it was going to land in the middle of town.'"

Many voiced their concerns about the missile range. The White Sands range was very close to populated areas. The Bumper Project, a second-stage rocket that had been added to the V-2, needed 75 miles more range. The White Sands range, 135 miles long, was too short for the newer and more powerful rockets. A longer test range was needed, and three potential sites emerged: one on the coast of northern Washington with a range along the Aleutian Islands of Alaska; a second at El Centro, California, with a range along the coast of Baja, Mexico; and a third at Banana River Naval Air Station, Florida, with launches from Cape Canaveral and a range over the Atlantic Ocean.

In 1947, the committee announced its decision for the proving ground at the California site, with Cape Canaveral offered as the second choice. Plans continued for the building of a range in California, but political problems arose in 1948. The site would be rejected when Mexican President Aleman refused to agree to allow rockets to fly over the Baja region . . . bad timing because of the V-2 that had recently crashed near Juarez, Mexico. Luckily the British would allow missiles to fly over the Bahamas and would also lease land to the USE for tracking stations. This sealed Cape Canaveral as the first U.S. Long Range Missile Proving Ground. On May 11, 1949, President Harry S. Truman signed legislation establishing "The Joint Long Range Proving Ground at Cape Canaveral." I would be graduating from high school the very next month.

On May 9, 1950, construction began on the first permanent access road and launch sites on Cape Canaveral. The first area developed for the launch operations became known as launch pads 1, 2, 3, and 4. On July 21, 1950, the Bahamian agreement with the British government allowed construction of the first tracking stations downrange. The range of tracking stations eventually stretched across the Atlantic Ocean to Ascension Island 5,000 miles downrange from Cape Canaveral. Later, some rockets went beyond Ascension, across the bottom tip of Africa, and into the Indian Ocean.

The Cape Canaveral Test Range was now in operation. The Army scheduled launches of two modified German V-2 rockets for July 1950. The rockets were called Bumper and employed a V-2 rocket as first stage and a WAC (Without Any Control) Corporal rocket as second stage. The first rocket launch at Cape Canaveral, that of Bumper #8, occurred on July 24, 1950. Cape Canaveral began to grow rapidly due to the testing of Air Force winged missiles and the introduction of ballistic missiles.

The biggest problem at this time was actually getting personnel to come work at the cape. The persons who wrote up job descriptions and employment advertisements were adding "Please come to Patrick." They were trying anything they could think of when writing job descriptions. Potential employees would show up; but after seeing the living conditions and what the area lacked, they decided they wouldn't and couldn't live there. It was very difficult to get required personnel, in part, because they had a hard time recruiting workers with families; and as a result, the cape was very slow in becoming properly manned.

On December 31, 1953, the Guided Missiles Range Division of Pan-American World Airways was granted the first range operation and maintenance contract. On February 28, 1954, Pan-American World Airways signed its own subcontract with RCA for the technical functions of operating and maintaining the range instrumentation systems.

This included flight data processing, electronic tracking instrumentation, photo tracking instrumentation, still and motion picture photo processing, and communication links between the launch sites and downrange tracking stations. RCA built a large photo lab at Patrick Air Force Base for all photo processing.

 Chapter 2

I was born on January 7, 1930, in St. Luke's Hospital in Utica, New York, to Genevieve (Jennie) Lewandowski and Howard Sherman Ehrenspeck, who were married on July 14, 1929. I was raised in a two-story house that was semi-attached to the Mandeville Meat Market owned and run by my grandfather, John Lewandowski, and my mother's brother, Henry Lewandowski. We lived in my grandparent's home, and my father supported us as a butcher at the market.

I have great childhood memories of helping in the market, learning how to make Polish sausage, sharpening knives, and growing vegetables in my own garden to sell in the market. My dad was a hands-on dad and taught me to fish, snow ski, ice skate, camp, and shoot a rifle. My mother recorded our adventures through photographs. I loved our family outings to Sylvan Beach, Niagara Falls, and other excursions with aunts, uncles, and cousins.

When WWII began in 1941, my father went to enlist but was turned away because he had a family . . . my dad's retelling, anyway. My uncle enlisted to take my father's place, leaving my dad to run the meat market in his absence.

Sometime in 1943 when I was around 13 years old, I was homebound for about a year due to rheumatic fever. I missed a year of public school; however, my grandmother and mother made sure my pals and cousins still came around to the house to hang out with me and my dog, Roscoe.

I think that it was during this time that I discovered I was very fond of building models, seeing how the pieces went together to make a finished product. My grandmother would make sure I always had models to build.

One day she brought a radio to me. She suggested I take it apart and put it back together. I remember my mother protesting, but grandmother said, "Eddie should try it! I know he will figure it out." And I did! I was very interested in how things worked and thus began my love of engineering.

My grandmother also brought a cookbook to me. When I was feeling better, we began on page one and she, my mother, and I cooked and baked everything in the cookbook, start to finish. Any overage would be sold in the market. I am grateful for this time we spent together. I was losing a year of school but was still using math, science, and reading with the activities with which I was involved.

In uniform at Manlius Military School

It was during this time I also discovered that I very much enjoyed playing the drums. My talent with them found me in a little, six-piece, local band.

The war ended; Uncle Hank came home and resumed working at the meat market with my dad.

School was very difficult for me. I was not happy nor was I successful. I heard about a military school, Manlius Military School. I went to my mom and dad; and, as was their habit, they listened patiently as I laid out the plan I envisioned would be my future.

My father contacted the school, and a representative came to our home to talk to my parents and then to me. At first, due to my previous health issues with rheumatic fever, they were inclined to say no to me. For some reason they eventually agreed, telling me that I would still be expected to hold my own even though they knew that physically I could not do everything expected. I was ecstatic. Thus began the best years of my life. I was even in the band; and yes, I played the stationary drums! I also played the piano, golfed, and met some guys who became lifelong friends of mine.

In 1951 or '52 my parents and I made the move to Hollywood, Florida. In addition to a short time working as a mail carrier and helping in the gift shop that my parents owned in 1953, I found myself at WITV, a local television station, working the cameras and learning as much as I could about the whole world that consisted of cameras, televisions, and photo equipment. I also became a second lieutenant in the Civil Air Patrol, as well as a HAM radio operator.

In July 1956, six years after the first launch at the cape and two years before NASA arrived at the cape, I was employed by the RCA Missile Test Project and started the most exciting, interesting—and at times, the most dangerous—employment of my life. My story is very similar to all the early space workers who believed in America and loved space exploration, yet initially lived and worked in less than comfortable surroundings. Their dedicated work ethics contributed tremendously to the advancements of today's space industry successes.

Cape Canaveral, meaning "thicket of cane," was aptly named. Surprisingly, it wasn't hard to believe that the United States government had picked a Godforsaken place like this to base a guided missile range. Some Army general working on the project probably joked about putting the site in the Florida swamps, and everyone thought he was serious . . . that seemed to be the way the government operated.

When I first saw the area, it looked as though very little had changed in the last 400 years since the 16th century when the Spanish began settling here. Outside of a few small towns and buildings, there was nothing but a narrow, two-lane, tarred road cutting through the dense scrub brush and palmetto trees. Tall, Australian pines shaded U.S. 1, the main route from Miami to Jacksonville. One could drive for miles without seeing any sign of civilization.

I later learned that the area just south of Cape Canaveral used to be one of the major training bases for the Navy during World War II. Seaplanes patrolled the Atlantic, searching for Nazi subs which many times managed to reach within a few hundred miles of Florida's central east coast. After the war, the Banana River Naval Air Station closed and would have never reopened if another proposed missile test center in California was not ruled out. A combined Army-Navy committee, selected by the Joint Chiefs of Staff, chose Cape Canaveral to build a long-range proving ground when political problems in Mexico prevented gaining approval to launch missiles from El Centro, California, downrange over the Gulf of California, which is bordered by Mexico on the east.

The newly-created Department of the Air Force took over custody of the air station in 1948; and, by the following year, plans already were being made for the first missile launch under the site's new designation: the Joint Long Range Proving Ground.

I vaguely remember hearing or reading something about that first launch on July 24, 1950. Not too much was said,

First launch at Cape Canaveral, July 24, 1950: Bumper 8

though, just that it was the first missile launched from the cape. As nearby Patrick Air Force Base grew and the public became more curious about what was going on a few miles north at the cape, the first launch, Bumper 8, took on a new significance like anything else which is usually first. People like to look back and recall how things all started. By the time I got there six years later, everyone was commenting how far we had come in so short a time . . . how it all began with Bumper 8, an Army WAC Corporal rocket mounted on a modified German V-2 missile similar to the V-1 buzz bombs Germany used to devastate England during the war. About 100 of the V-2 missiles were brought to the United States for study after the war, so it was only logical that these would be used as our first attempts at building a solid military defense system.

In early 1956, advertisements for employment at Cape Canaveral were very numerous and were found in newspapers, magazines, and other various print media. I was employed by the telephone company in Fort Lauderdale, Florida, at the time.

I had always been interested in rockets, and there appeared a slim chance I could be working near the cape where I could watch and learn more about missiles: unbeknown to me, in February of that year, my wife, Jan, sent in a coupon for information about working at the cape. The advertisement appeared in some monthly electronics magazine I had purchased. I always read trade journals as they helped me in my job as a frame man for the phone company. I mentioned to Jan it would be great to work at the cape, showing her an ad. I also had experience working as an audio engineer for RCA Victor out in California for a while after we were married, but I didn't believe I had the kind of knowledge or the skills for the type of work working at the cape would require.

I soon forgot my dreams and the ad and tried to make the best out of working for the phone company at $46 a week, paying for a house and supporting a wife and a three- month-old baby. We never even had enough money to go out to eat much; so, every two weeks when I got paid, I would stop on the way home from work and buy a six-pack of hamburgers at Royal Castle. That was our big night. We would sit in front of the TV, eating our dinner of tiny hamburgers together.

——————————————————————— Chapter 3

One afternoon I came home from work and picked up the mail off the kitchen table on my way through.

"What's this?" I asked Jan.

"What?"

"This letter. It's from the United States government."

"Open it and find out."

She stopped making dinner and stood next to me, watching.

"Melbourne, Florida. There's nothing up there."

"No?" Jan was smiling as though she knew what it was.

I carefully opened the envelope and pulled out the papers inside.

"It's from RCA, Missile Test Project."

"Really? What do they want?"

"There's a letter here asking me to fill out this application. Look at this thing! I wonder where they got my name."

"I sent it in."

"When?"

"About two weeks ago . . . when you showed me that ad. I figured if you weren't going to mail it, then I would."

"That's great, but . . ."

"But what? What's wrong?"

"I can't go to work up there. You know that."

"And why not?"

"What would I do? I don't know the first thing about missiles. I don't even know what the heck is going on up there."

"You mean you're not even going to fill out the application? Why not? You have just as much experience as anyone else.

Look, they said in the ad they needed people with some experience in electronics."

"But all I can do is fix TV sets. I can't fix missiles!"

"You could learn."

"I don't know . . ."

"Do you want to stay with the phone company the rest of your life? Don't let me ever hear you complain about it again, though."

"Alright! I'll fill it out. I guess it wouldn't hurt to try. Who knows? Maybe I could get a job there. That would be something--working around the missiles. It sure would be a lot more interesting than playing with phone circuits all day. I wonder what they pay. Has to be more than what I'm getting now."

That night during dinner we talked and dreamed about what it would be like living and working at the cape: the missiles; the prestige; being able to buy all the things we always wanted but, like most other people, couldn't afford. What more could I ask for? Those Buck Rogers serials which fascinated me so much when I was a kid were now becoming more real in my mind and in my life. There was a *chance*. All I had to do now was send in the application. We filled it out after dinner and sent it off the next morning.

Each day I thought more and more about the cape: working there and watching the missiles. That seemed to help me make it through the day. Splicing and disconnecting phone cable all day presented no challenge to me, mentally or physically. It was just a job . . . and boring at that. I knew there was a good chance I wouldn't be qualified to work at the cape, but I would be happy to do anything—wash windows or cut the grass— just to be around the area and be a part of it. Perhaps in a subconscious way, I realized I didn't have the background to work at the cape in any technical position, and I was preparing myself for the worst. It would be too disappointing if I didn't get the job. If I kept thinking I wouldn't get the job and I didn't, then I wouldn't lose anything. However, I hoped I had some chance.

When another letter came from RCA, I was afraid to open it. Even though there was a chance—a slight possibility—that I would be accepted, I knew what the realities were. Slowly I opened the white envelope, trying to prepare myself for what I almost knew the letter would say. My weeks of hoping and dreaming would now end. I wished I had never sent in the application. How pompous of me!

As I began reading, my eyes seemed to hang on every word, pulling as much meaning as I could from each sentence.

"They want me! They want me!" I shouted as I plopped down in the nearest chair in disbelief. "They want me to come up for an interview . . . at my convenience."

Finally, I could relax.

I started reading the letter out loud to Jan:

"Dear Mr. Ehrenspeck:

Thank you for your recent interest in RCA and its operations in Cocoa Beach, Florida, at Cape Canaveral. As you know, we are working with the United States Air Force at the missile test center there.

In the past several weeks, we have reviewed several thousand applications for employment, and we apologize for any delay in responding to your initial query. After careful consideration, we have concluded that you may be qualified for one or more of the several functions being performed by RCA at Cape Canaveral. We were extremely impressed with your background and previous work experience.

We would very much like the opportunity to meet with you to discuss in further detail your application for employment. Please contact the office below to arrange an appointment at your earliest convenience.

Thank you for your interest in RCA, and we look forward to your reply.

> *Sincerely,*
> *Robert A. Smith*
> *Operations Manager,*
> *RCA Missile Test Project*
> *Melbourne, Florida"*

"Can you believe it? They want me to come up for an interview at my convenience. I can't 'believe it! I'm glad you sent in that coupon. I can see it all now. What do you think I'll be doing up there?"

"Hold on! You don't have the job yet. Go up for the interview first. You may not like what they have open," Jan answered.

"Are you kidding? Not like being around missiles? How could I not like it? Look at this letter!" I exclaimed, starting to read it over again. "I still can't believe it!"

I wrote back immediately, telling them I would be up the following Saturday for an interview.

All the way there I kept picturing in my mind, over and over, the tiny, foot-high missile I built as a kid. You couldn't even buy rocket models back in the 30s. I designed it myself from drawings I had seen in comic books and at the movies. I cut out strips of thin balsa wood in the shape of fins and glued three of them to a crude, cylindrical body. It wore out within a week from constantly carrying it everywhere I went.

When we finally arrived in Melbourne four hours from our house in Hollywood, Jan, along with our baby, waited outside in the car while I went in for the interview. The building RCA occupied looked lost next to the three-story New Haven Hotel. A few hundred feet away to the east was the Intracoastal Waterway. I could clearly see the entire two-lane, wooden bridge which connected the main part of Melbourne to the beach. From a distance, the cars seemed to touch when they passed.

The day felt perfect. Never had any place I ever visited looked so beautiful. Even the trees looked greener up here. At the time, I was so caught up in the possibility that I soon might be working at the cape that I probably saw what I had hoped I would see . . . not what was actually there.

I went inside, introduced myself, and sat down along side of the desk where a Mr. Turner motioned me to sit. The place was nothing fancy . . . just a big, empty room with wooden desks that had pencils on them. The tall windows on both sides of the room made me think of an old schoolhouse.

Turner and I talked for a few minutes about how I had found out about the opening at the cape and where I had worked before. He looked familiar to me, but I knew there wasn't any way I could have ever met him. He said he had just moved down from Chicago. Maybe it was the way he looked at me, pretending to be interested in what I had to say . . . like most other people who conduct interviews.

After the formalities were out of the way and Turner determined I was halfway normal, he opened one of the lower desk drawers and pulled out a white booklet.

"In order for us to proceed," he explained, "you're going to have to take an exam."

Immediately I froze inside. I was never any good on tests. Plus, I never expected this. I thought about what I should and shouldn't say during the interview and how I should act, but I never prepared myself mentally for any kind of test.

"Well, what does it consist of?" I asked, trying to push all thoughts of failure further back in my mind. I knew he could see I was upset from the scared look that suddenly appeared on my face. That made it worse.

"It's all multiple choice. Do what you can," Turner said in a calm, reassuring voice. "There's some writing in there, too, so you can tell us what you know about certain equipment."

He lifted the test toward me and announced it would be okay to sit anywhere.

I stood next to his desk for a few seconds, thumbing through the pages. It was almost an inch thick! Hoping there was some way out, I told him I didn't know too much about some of the technical equipment covered in the last part of the exam.

"Well, take the parts you know. You've got to have a score. I know that from talking to you and from your background and experience, there should be no problem. You should be able to take this test and pass it with no problem at all," Turner repeated.

"I'll look through it," I offered.

"Look through it, and do what you can," Turner pushed.

That sounded like a nice way of telling me to quit complaining and start working on the damn test, that he didn't want to be bothered again. He probably wanted to leave early so he could stop and have a beer with the boys on his way home.

I began reading some of the questions. They wanted to find out how much I knew about radar, telemetry, high-frequency communications, electronic diagramming, and three more hours' worth of other things undoubtedly much harder. If I had been in the service and studied some of this, I might have been able to do all right . . . or at least answer some of the questions. However, there was no way I could even hope to come close to passing . . . no way in hell! All I could do was look at the test!

In the five minutes I sat there, all I could think about was Jan out in the car with our daughter, Jennifer, waiting and hoping everything would go okay for me. Where would I find the words to tell her that our dream for the past month was only that: a long, lost dream? Now it was really over. All the hope I had for the future died.

I couldn't just sit there and let this chance go by. I had to do something. Without mapping my strategy, I grabbed the booklet, marched back up to the front of the room, and threw it on Turner's desk. The test snapped when it hit the hard wooden top.

"No way," I announced. "If you want me to come and work here and do a good job and learn anything you can teach me and be here until this place closes, you've got me! But if everything depends on this test, you don't have me!"

I couldn't believe what I had just done. I was so nervous I was rambling. Now I was worse off than before. I just quit a job I didn't even have. Maybe I could plead insanity.

"You haven't even put your name on it," Turner observed.

"What for? I can't even get through the first page. You know what I know, and I know what I know. I know what my learning capabilities are and what I can do, but I couldn't pass this test right now if my life depended on it."

He picked up the booklet, threw it back in the same drawer, and said, "Forget the test. Let's start talking about going to work."

Wow! Five, ten minutes ago I had literally died; and now I was reborn! It was a miracle! Turner was an all right guy . . . hard to predict, but all right. Once again, I had a dream. I knew I would always be grateful to Turner.

Chapter 4

The following week, Jan and I talked almost constantly about the cape and our new life. Although I was now making $52 a week at the phone company, it seemed very little compared to the $60 per week I would soon be earning and the opportunity waiting for us just 200 miles further north.

Next Saturday, we went back up for another interview. This time I had to go to Patrick Air Force Base to meet the manager and some of the people with whom I would be working. Again, Jan and the baby waited outside.

I learned that RCA scheduled me to work in data reduction. My job would be to read 35mm film taken of the launches and record specific information. One of the people with whom I spoke at the base said they had talked with Turner and added that it appeared as though I would be able to make it. In less than half an hour the second interview was over. One of the managers said they would get back to me in a few days for "finalizing"—whatever that meant—and when I was to begin work.

I had the job for sure. Unless something unexpected happened, I would be working at the cape! I wouldn't be directly onsite, but I would be at the cape. That was all I wanted originally.

On the way home, I started thinking . . . thinking that I wouldn't be able to see the missiles or watch the launches. A few weeks ago, I would have been happy doing anything just to be near the cape. Now, I was thinking differently. I wanted more. I wanted to be involved in more than sitting in a room somewhere, miles away, looking at film. Better not say anything now, though; be thankful and leave it at that. For some reason—greed perhaps—my mind was months ahead. I hoped that maybe later, when I was there a year or two, I could push to get into something else. Wait till then.

I tried to reach Mr. Turner to inquire as to when I might actually start work. The secretary said he was gone for the day and wouldn't return until next Monday. I couldn't wait that long. She transferred my call to Patrick Air Force Base. It rang at least 10 times before someone answered.

"Hello. RCA Data Reduction Center. Can I help you?"

"Uh, yes, ma'am. Is the manager in, please?"

"Yes, I believe he is. Can I tell him who's calling?"

"This is Edward Ehrenspeck."

"Hare-in-deck?"

"No, Ehrenspeck. E-H-R-E-N-S-P-E-C-K. Edward Ehrenspeck. Can you hurry? This is long distance."

I knew if I ran out of dimes, I would have to go back to the apartment, get more money, and then find someplace to get change. By then the office would probably be closed, and I would have to call back tomorrow.

"Hold on a minute, sir."

I could hear her ask another person nearby if Norm was still there.

"I'm sorry, sir, but Mr. Poole just left. Can I take a message, or is there anything I can help you with?"

"Well, uh, I'm supposed to start working in data reduction soon; and I was wondering when I was scheduled to start. They told me they would contact me, but it's been about a month; and I still haven't heard from anyone. Do you know anything about it?"

"Let me see what I can do. Let me see if I can find your file. It should be here somewhere if you were assigned to this department. Okay, let's see . . .," the lady drawled out. "Eady, Eason, East, Eckert, Edelmann, Edwards, Egan, Eggers, Ehley, Ehret, Eldon . . .," she read aloud as I heard her fumbling through files. "Sorry, sir, nothing. You did give me the correct spelling?" she suddenly interjected.

Great! I'm running out of change, and some idiot wants to know if I can spell my own name.

"Yes. E-H-R-E-N-S-P-E-C-K. The first name is Edward."

"Sorry, but I don't have anything listed under that name," the lady managed to offer.

"It's got to be there! I had an interview with the manager last month, and he said it was all arranged for me to work there. Is there anyone else there who could help me? Are there any other files you could look through? Maybe my folder was misplaced. It has to be there somewhere," I lamented.

"Let me look on Mr. Poole's desk. Perhaps he had it out of the file for some reason." She put me on hold and walked to Poole's office. "Nope, it's not here, either," she spoke as she picked up the phone in Poole's office to continue our conversation. "I don't know what to tell you. Why don't you call back in the morning? Mr. Poole usually gets in around 10 a.m. I'm sure he'll be able to help you."

"Okay. Thank you," I replied disappointedly.

What was going on? It would have been almost impossible to lose the entire file. Even if they did, how could they forget they had an interview with me only four weeks ago?

On the way back to my apartment, I decided not to tell Jan about what happened. There was no need for her to start worrying and get all upset. I would tell her they told me to call back tomorrow morning. The manager would be in then.

As soon as I walked in the front door, Jan handed me a letter which came while I was out. It was special delivery from Cocoa.

"You are to report to Patrick Air Force Base, Building 45, for orientation and training at 9 a.m., July 16."

That was in only four days.

The letter went on to say I was to go to a certain motel in Cocoa Beach where RCA would pay all my expenses for the first 45 days. In addition, I was "hereby advised" that because of my past experience in electronics and my interest in photography, I would be working as a camera operator instead of a film reader in data reduction. I was actually going to film the launches! I would be able to see, hear, and feel the missiles . . . be close to them! We began packing immediately.

I'll never forget my supervisor at Southern Bell when I went in the next morning and said it was my last day.

"If you quit me now," he warned, "you'll never get another job with the phone company . . . *ever*!"

He had to be kidding! Why the hell would I even want or need to go back to the phone company after working at the cape? He could keep his $52 a week . . . hire some other fool to work there.

We made plans to move Saturday. I notified the landlord and went down the street to rent a U-Haul trailer. It was an open trailer, the cheapest they had . . . small but large enough to carry the few things we owned and were taking with us.

I packed the trunk of the car—a '55 Chevy—early that Saturday and put the rest of our things in the trailer. I secured the canvas cover just in case it rained somewhere along the way. In Florida, it could be clear one minute, rain 10 minutes later, then be sunny again.

We left a little after noon, hoping to get to Cocoa Beach by dinner time. Because of the trailer, I had to drive about 10 miles an hour slower than usual. The normal speed on U.S. 1 was 45 but only 35 when passing through a town. Considering that—plus tourists and regular weekend traffic—I figured it would take about five or six hours to get there.

With less than $200 to our name and a trailer stuffed with household goods we either couldn't sell or we actually needed, we left South Florida and everyone we knew. It was a lonely feeling. The most important things we brought with us belonged to the baby: crib, bottle sterilizer, playpen, and her clothes. In the back seat was our black and white television set which wasn't paid off yet. Fourteen more payments, and it would be ours.

We were excited about going to the cape despite feeling somewhat leery. This was all new—not only to us but to the rest of the country. I wondered how long it would last and whether I had made the right decision. We weren't really sure what we were getting into. Thinking it out, we rationalized that with a company like RCA and adding in the United States Air Force—both for which I would be working as a civilian—we had nothing to worry about. Everything would be taken care of . . . hopefully!

Driving there, I kept seeing visions of Flash Gordon and his spaceship . . . a dream come true. I was sure Jan was thinking about all the things we could finally afford once I started working.

"Anything wrong?" I asked about halfway there.

"No. Not really. Why?"

"I don't know. You don't look happy."

"I am. Just thinking, I guess."

"About what?"

"Oh, leaving Fort Lauderdale. I'm happy to get out of there. Maybe this place will be a little better."

"What do you mean?"

"Well, for one, your parents won't be there."

"We went through all that before. Do you have to bring it up again?"

"Well, you asked what was wrong, and I told you. You know, I can't understand why parents always interfere with their children's lives, even after they get married."

"I don't know; that's the way parents are. We'll probably do the same thing when Jennifer grows up."

"Not me. Once our children get married and move away, that's it."

"You'll change," I challenged Jan.

"You might, but not me."

"What else is bothering you?" I pressed.

"I was just thinking how good it feels to get out of south Florida even though all our friends are down there. I never really did like it down there. Probably never would."

"Why not?"

"No one seems to care much about anyone else. It certainly wasn't like that back home in Ohio. Besides, you can't make any money down there."

"We will now! Sixty bucks a week . . . just to start! Think of what I could be making five years from now! If I play my cards right, I could be pretty well along by then . . . maybe a manager or a supervisor. Not bad for being only 26 years old! We gonna do alright," I said emphatically as I looked over at her and winked.

I was lucky, and I knew it! Not everyone was able to reach their dreams . . . especially back then. Jobs were hard to find, and there were people everywhere looking for work. I not only had a job, but I was also getting into a field I thoroughly enjoyed.

That was a lot more than most people had. Even those who were fortunate enough to be employed rarely found it in the field of their choice.

 Chapter 5

As we approached Melbourne, it was half past five and raining. We took the first causeway over to the beach and began looking for our motel. Everything had been arranged; all I had to do was get there. With RCA picking up the tab, we were traveling first class. They were even paying for the gas: 10 cents a mile.

It was hard to see anything through the front window between the rain and the windshield wipers slopping back and forth. It seemed like we had been driving for miles after turning onto the causeway and would soon run out of highway. We were getting hungry, and so was the baby. I hadn't eaten anything since breakfast. Jennifer had been awake for the past hour, and it was only a matter of time before she would start screaming for her bottle.

We didn't know where we were; and, worst of all, we had no idea where we were going. We felt alone and uneasy; nothing we passed looked familiar. Suddenly, we saw something just ahead on the right . . . a red, flashing sign. It was the motel!

Someone in the office saw us pull in and ran outside before I had a chance to get out. A man holding a newspaper over his head went to the closest side of the car. Jan quickly rolled down her window.

"Sorry, we're all filled up here," he said. "We haven't got anything open."

"Wait a minute!" I shouted above the rain. "Hold it!" I added for emphasis. "I'm gonna be working at the cape . . . with RCA. They told me to stop by here, and this would be the place you'd have a room for us."

"Sorry, we don't have any rooms. Glad to meet you, though. I work for RCA, too!"

I felt a little better meeting someone else who was working at the cape. Still, I kept worrying where we would spend tonight.

"What am I going to do?" I persisted. "Any suggestions?"

"Gee, I don't know. There aren't many places around here to start with, and I'm sure every place is jammed up for the night."

"I'm going to have to do something. I've got a baby here, and we have to find a place . . ."

The man interrupted me, "Let me make a phone call. I think I know where there might be a spot. We have some people who come here on business and don't stay as long as they planned. If you catch such an opening, you've got it made. Hold on!"

The man ran inside and returned in less than a minute.

"Hey, down there at the Brevard Hotel, they got a room for you. But man, you'd better put that thing in second gear and get over there! Go down to the next signal and take a left. As soon as you get over the bridge, you'll see it on your left. Good luck!"

"Thanks a lot!" I said with heartfelt gratitude. "Hope to see you again someday."

I shot out of the parking lot and headed for the causeway. Just on the other side of the Banana and Indian Rivers in downtown Cocoa was our temporary home. We had finally made it.

Everything looked much better the next day, Sunday. The rain had ended, and the sun was shining. We spent most of the day unpacking and fixing up our provisional home in the Brevard Hotel.

The room was small and overlooked a short, wooden fishing pier on the western edge of the Indian River. The full-sized bed was pushed against one wall, and a dresser stood along the opposite wall at the bed's foot. As with most hotel rooms, there was a chair and a table next to the window; and a nightstand stood about four steps away. The room included a small kitchenette. Everything had been positioned so one could walk around the room without bumping into anything.

There was even less room after we set up our television, Jennifer's crib and playpen, and brought in all of our clothes from the car.

As for baby formula, the kitchenette had a very small refrigerator. We were able to prepare the formula but not able to refrigerate it properly. The hotel kitchen, which was always open, had an excellent refrigeration area; and sections were prepared for baby formulas to be refrigerated. Names of all infants were put on the bottles by the parents.

This was class: the Brevard Hotel. It was one of the first hotels in the area which years ago catered only to well-to-do visitors from the north who came down during the winter months. Every part of the hotel had a quiet elegance that spoke of better, earlier times. Built in the early 1920s, the hotel was known as "The Grand Old Lady." During those times we wouldn't have been able to afford a room there for even just one night. It hosted many dignitaries, including Vice President Hubert Humphrey. Even now, some 10 or 20 years later, the place was clean and well-kept. In 1996, the property would be sold and developed into condominiums.

For now, however, the coolest spot in the entire building was out back on the wide veranda. Many nights I would have liked to sleep there instead of sweltering in our room. It was relaxing to sit there on the weekends and just stare at the river floating lazily by. Occasionally, a boat or two would pass. This was quite a contrast from Fort Lauderdale. The whole area was much less congested: no crowds, concrete high rises, or miles of asphalt leading off in a million directions. Yes, this was a refreshing change.

The very next day I was to sign in at Patrick Air Force Base, 15 miles south of the cape itself, as a new employee for the RCA Missile Test Project. I was to report to Building 45 at 0900 hours for examination and training. Although I wouldn't actually be working for the Air Force, Patrick was the only place large enough with the facilities to train all the people who were now being hired to work at the cape.

RCA and Pan American Airways had several contracts with the government to provide certain services at the cape. A proposal was made in 1952 that management and operations be transferred from the Air Force to a civilian contractor because of periodic turnover of service personnel. Pan Am was informed of the decision immediately because of its past experience in building air bases around the world; and one year later, after considering dozens of other firms, the Air Force announced that Pan Am would be the range contractor and RCA the subcontractor. Pan Am would provide launch support and maintenance service while RCA agreed to develop a system to record data, such as the pictures I would be taking of the missiles. Hundreds of people like me were now being hired to keep up with the program. Only 65 people were working for Pan Am in December of 1953, and by 1957 there were almost 4,000! That was just Pan Am. There also were service personnel at Patrick Air Force Base and people at RCA and several other smaller companies.

I spent my first two days mostly waiting in long lines, being herded in and out of buildings for various tests and examinations. There were all types of people waiting for physicals: men, women, engineers, instructors, electronics specialists, scientists. It was like the Army: we waited outside in the sweltering summer sun until our number was called, and then they directed us to various stations inside the wooden barracks.

Looking up and down the lines, I realized how funny we all must have looked, standing there holding little, plastic, urine bottles. The paperwork was also enormous.

After my physical, they informed me I would be assigned to the Cape Canaveral Theodolite Section. They handed me the schedules of about 13 different schools I would be attending, schools that taught courses ranging from Cine theodolite operation and maintenance, photography, communications, timing, astronomy, tracking, cape operations and familiarization, and the list continued. Mostly, I would learn how to operate the camera I would be using.

For the next three months, from mid-July to September, I memorized everything there was to know about the theodolite. I spent eight hours every day, five days each week in a stuffy classroom and studied three to four hours at night, as well, when I got back to the hotel. Jan always went out fishing on the pier in back so she wouldn't disturb me. She spent most of the weekends alone while I studied.

As the course continued, I found I had to spend more time studying in order to keep up with everyone else in class. My knowledge of electronics was extremely limited, and now I was being hit with things I had never even heard of before: binary codes, flip- flops, inversions. I could fix a simple black-and-white TV set, but this was 10 times harder! When I first saw pictures of the theodolite, I was excited and anxious to learn all about it and how it worked. Now that the enthusiasm was wearing off, many times I was thinking there was something to the adage, "You can't teach an old dog new tricks." I was an old dog trying to learn something new. Studying was for kids in high school, not me. There was no way out, though; I had to study. There was too much at stake now: if I didn't pass, I couldn't even go back to the phone company.

One afternoon a week or two after I started school, I heard a few guys in class talking and found out they were making $70 a week, $10 more than me. I checked around and learned that some of them didn't have nearly the experience or knowledge that I did; a few didn't even know how to work a regular 35mm camera. So why was I making less? Before I went back to class after lunch, I decided to talk with someone upstairs where my file was kept.

"Excuse me. I'd like to talk to the manager," I told the girl at the front desk in the personnel office. "Is that Mr. Simpson?"

"Yes, it is," she responded. "Is he expecting you?"

"Uh, no. Tell him Edward Ehrenspeck is here. I'm in the theodolite school downstairs."

"Sure. Let me check with him. I'll be right back."

While I stood there waiting, I tried to figure out exactly what I should say to Mr. Simpson. After all, I hadn't even been here a month; and here I was asking for more money. I deserved as much as everyone else, though, I told myself. Why shouldn't I get the other $10? It wasn't the money as much as it was the principle. I was just as good as anyone else in the section.

"Go on in," the young lady jerked me out of my reverie. "Mr. Simpson will be happy to see you."

"Thank you!"

I walked into his office and stopped as soon as I looked at

the man sitting behind the desk. It was the same man I met the first night when we were looking for a room in Cocoa Beach.

"Mr. Simpson? Jack Simpson? I'm Ed Ehrenspeck. Remember me? I met you at the motel when we first pulled into town."

"Sure, sure! Come on in. Sit down. How have you been? Make it okay over to the Brevard?"

"I sure did! I want to thank you for helping me out. I don't know what I would have done if you hadn't called over there."

"No need to thank me. It was the least I could do. It's pretty rough trying to find a place to stay around here. There can't be more than 10 places in the entire area. Well, what do you think so far? What outfit are you with?"

"Theodolites. Things are going pretty well. I hope I can stick with it. I never had too much training in electronics, but I'm learning . . . just takes time. The reason I'm here is to find out about something."

"What's that?"

"Well, I understand that most of the other guys in my class are making $70 a week. I'm only getting $60. Is there any reason why I'm getting less? I believe I have just as much experience as most of them."

"Let me have a look at your application," Simpson offered.

He wrote down the spelling of my last name as I gave it to him and had the secretary bring in the file.

"Okay. You have experience in radio and TV repair; you've worked as an audio engineer at a TV station; you were a TV technician with RCA out in California and most recently a frame man for the phone company . . . I don't see any problem here. You've got it!"

"I do?" I choked out in astonishment. "Just like that?"

"Yep. Someone must have made a mistake when they hired you. The rate for camera operators/technicians is $70 a week. I'll take care of it."

"Hey, thanks a lot! I really appreciate it."

"No problem. You should get as much as everyone else. Glad to help you out."

"Well, take it easy," I offered as I got ready to exit the office. "Again, thanks."

"Okay. Good luck. If there's ever anything else I can help you with, just let me know. Keep in touch."

I wanted to run down the hall! I had been there only two weeks, and I had already received a $10 raise! RCA was doing things right. They knew how to run a company. I supposed they were helping all their people because of the contract they had with the Air Force. If they didn't come through, they wouldn't get paid; so, they took care of us.

_____ Chapter 6

RCA set up a Missile Test Project Club, and they gave us a club card and a sheet explaining all our benefits. We received a sticker for the car; they gave us a booklet of rules and regulations concerning the base (where we could go and what was off limits); and we also had a card which was good for discounts at various businesses in the area . . . not bad benefits.

During the day while I was at school, Jan mostly cleaned up the room and took care of Jennifer. For lunch she usually walked into town, pushing the baby in her stroller, along with a few of the other wives whose husbands worked at the cape. At night we had to go back into town to eat because for some reason the dining room at the hotel was closed that time of year. Every morning we ate breakfast in the hotel lobby where the staff laid out coffee and cinnamon rolls on a card table. We saw a lot of each other: everyone was always eating together. With only one restaurant near the hotel and breakfast in the lobby, we saw the other people who were staying there almost daily.

After a few weeks, life became routine. It was a slow, confusing time for us. Maybe my father was right: what were we doing here after all? All I did was study. Jan and I never went anywhere or did anything together except eat breakfast and dinner. This wasn't the way we pictured it. There was no air conditioning in the hotel room or at school, and the days were getting hotter . . . sometimes 95 degrees. We had to find an apartment of our own by the end of August or rent the room in which we were staying, which would cost more than I made each week. I was getting tired. Things didn't seem to be getting any better. I had been hired to work at the cape, and I hadn't even been there once. For all I knew, the cape wasn't even built yet.

Many times in those first few weeks, I wondered if I had made the right decision . . . to leave a steady job for something like this. No matter what, I couldn't go back there now. I felt I had something to prove not only to myself but also to the people back home who thought I wasn't smart enough to work at the cape and who believed I was insane for leaving a good job, one with a future and chance to move up. I had to show them I was right. That was the only thing that kept me going.

We felt a little better once we moved into our own apartment. Our stay at the Cocoa Hotel was provided by RCA for 45 days. During that time, we were to find other housing for the future; and all living responsibilities from that time forward were to be paid by us. We got word that a string of small duplexes had been built and were now available for renting. They were in the city of Cape Canaveral right on the beach. We investigated the duplexes immediately and were fortunate to be able to rent one of them. However, it wasn't much better than the hotel room; but at least this place was ours, and we had a sense of security. We no longer thought of ourselves as visitors. This was our home now. The lease was for a year, and we would be there at least that long no matter what happened.

Our two-bedroom apartment was about 200 yards from the ocean. There were several concrete block triplexes, all side by side, facing the beach. We lived in No. 55, between two other apartments. Each of the buildings were painted bright yellow, inside and out, and the trim around the doors and windows was different on each: brown, pink, or turquoise.

We were one of the first tenants to move into the fresh, new Winslow Beach Garden Apartments in Cocoa Beach. The people who ended up moving in on both sides of us and in the other triplexes were pleasant, most of them working at the cape, also.

The rent was $100 a month, unfurnished, no air conditioning. When the ad said "unfurnished," they meant it: a stove and refrigerator would have cost $25 more per month;

there weren't even any towel racks or a toilet paper holder in the bathroom. The floor was concrete; and a few months after we lived there, we took everything out so they could paint it green. The only things we had when we moved in were the baby's furniture, a dresser, and a bed we bought on time. The place wasn't big enough for much else. Both bedrooms together would have made a fair-sized master bedroom, and the bathroom didn't even have a bathtub . . . there wasn't room enough.

As small as it was, the apartment still looked empty. After we moved in, there was nothing in the living room and kitchen except for several boxes of linens, dishes, clothes, and two folding chairs which had been in storage the past month while we were living in the hotel. We had a large ice chest that we immediately filled with ice for refrigeration.

We had received a large fry pan as a wedding gift that would bake cakes, roast meat, fry foods, cook stews, and more. We cooked on a $1.99 barbecue grill out back on top of a garbage can until we could afford a two-burner hotplate. We now had a "kitchen." There were no phone lines in the duplex, no ceiling fans, and no air-conditioning; but we were fortunate that we at least had a small, gas space heater in the wall and a hot water heater.

While attending classes at Patrick Air Force Base, I was fortunate to make friends with a young man who had moved from another state to work at the cape. During one of our conversations, he mentioned that he had moved his furniture and appliances to Florida and had them stored in Cocoa. He was perfectly satisfied with his living conditions, a small room rented in Cocoa Beach; and he was about to start selling his belongings in storage due to the high cost of storage. As a result of our conversation, I was able to purchase a small refrigerator and stove for a giveaway price. We had brought two umbrella chairs along with two or three lamps with us from Fort Lauderdale but had no tables. One of our new friends gave us a card table with some plastic chairs and a rocker with the wooden rockers sawed off. We went to Sears in Orlando one weekend, about 100 miles away, and bought a three-piece set of lawn furniture for the living room. Between two of the pieces, a brown, cardboard box served as a corner table. We put a small Christmas tree on it that year. We were now settled in and on our way.

Water from the well below always smelled. Since the pump operated on electricity, there were numerous times each week when the power went out; and we couldn't flush the toilet or get a drink. In those days the power company wasn't too reliable.

A few doors down lived a couple who came from Boston. Somehow, with four children, they fit into the same size apartment as us. The company for which he worked at the cape paid their rent; and he also earned a regular salary plus "swamp pay," which was additional compensation for having to work in that area. We became good friends in the year he was there, and he tried several times to convince me to move up to Boston where he was studying astrophysics. He was working on some type of contraption known as the B.U. Telescope. I didn't know too much about it then. Years later I would use it quite often.

Shopping for anything was difficult. There weren't many stores, and the few that were in Melbourne or Cocoa usually didn't carry very much. A place called Storekeeper by the Sea was inside the nearby Jake's Bowling Alley at the corner of U.S. 520 and A1A on the beach. It was convenient for milk, bread, and cigarettes. For meat and vegetables there were a Winn Dixie and A & P in downtown Cocoa. Once a month we would go to Publix in Orlando and stock up on canned goods.

There were two or three clothing stores in Cocoa and a five and dime. For several years there weren't any other chain stores in the area. Other than these few stores, there just wasn't anywhere else to shop.

Even the availability of Sunday church services became a problem when every church would be filled to capacity and people waited outside for the next service. Jake's Bowling Alley even graciously opened on Sunday morning to allow services to be held there. The religious altars were set up on the alley, and visitors would sit where they would normally sit to watch bowling. Once one denomination finished its services, another would set up. This helped alleviate the lack of churches.

The townspeople, not only in Cocoa but also in Melbourne, for many years later resented the fact that suddenly thousands of people were moving into the area, demanding all sorts of things that the longtime residents never even heard of. The area wasn't prepared for what was happening at the cape.

Roads were bad, stores were few, and there weren't many places to live. People who had rooms in their houses were renting them out; and even those who had money, lived in shacks and trailers on Merritt Island because there was no place else to go. Some people even slept in large sewer pipes which had not yet been put in the ground. Along Route 3, running north and south on Merritt Island, there couldn't have been more than 20 or 30 actual houses from the south end to the north end near Titusville, a 25-mile stretch. Their sleepy town was rapidly growing into a full-fledged city, and we were the cause. The county's economy was no longer dependent on the orange groves as in the past. As more people moved in, construction flourished; and new cities went up almost overnight.

Satellite Beach was one of about four new communities being built in 1956 to accommodate the thousands who were being hired by the Air Force, RCA, Pan Am, and other companies at the cape. Percy Hedgecock, a fairly wealthy businessman, was the developer of Satellite Beach. He designed it, financed most of the construction, and later became its mayor. For many months after we first moved to the area, Satellite Beach had the only gas station on the beach. One always needed to make sure they had enough gas if they were going over to the beach from Melbourne or Cocoa unless they were going by the Gulf station.

Like everything else, living conditions in the area got much worse before it got better. More and more people came in every day.

Chapter 7

I finished school a few weeks after we moved into the apartment. My test scores were good enough to pass and to give me the opportunity to actually operate the theodolite. Classroom training gave us the knowledge we needed to understand the mechanics and electronics end of the camera, but now we had to put that knowledge into practice.

The camera itself was easy to use compared to the instruments which recorded pictures of the missiles, giving an elevation and angle reading on the film as the launch was tracked downrange. The film was later analyzed back at the base by the film readers.

From the classroom we were sent to the optics shop where we cleaned the equipment, loaded cameras on and off the trucks, and did minor repairs and other maintenance. Learning some of the problems and how to solve them helped us once we were out in the field, miles from everything, filming the missiles.

We were unable to be sent to the cape until installation of our new tracking equipment was completed. Until then we were to remain at Patrick to practice and hone our tracking skills. To help us become more familiar with operating the cameras, each day we would take the equipment from Patrick Air Force Base to the beach across the street and track Air Force planes taking off, landing, and flying up and down the coast. Even though we were on the beach where there was always a breeze coming off the ocean, I'll never forget how hot it got. Sweat ran down my forehead into my eyes so badly that

I couldn't track some of the planes. The classrooms were cool compared to it.

About two weeks later we received orders to be sent to the cape. Our new equipment had been installed and was in excellent working condition. We were given a date and time to be at Patrick and then transported to our tracking section at the cape.

We boarded a bus for our first look at the cape. It took us nearly 45 minutes to get from the base to the central part of the cape. We rode for miles, seeing only barren wilderness and dusty roads leading off to nowhere. I had no idea as I rode along that once I was transferred to the cape, I would be out there in the middle of it all without any protection, filming launches.

No one on the bus had been to the cape before, and we didn't know quite what to expect. I kept thinking of Flash Gordon, master control, and rockets lined up in rows. Surely there would be one or two launches while we were there. But all I saw was land . . . empty, flat land. Off in the distance I could see a gantry, the framework that supported a rocket prior to launching, but no rockets.

The buses drove down through the industrial area which consisted of a few small buildings and a hangar. Just inside Hangar C, I caught a glimpse of a sleek, white missile lying on a platform. I wasn't the only one who had seen it; everyone else also turned around, craning their necks, as the bus continued down the road toward the Optics Building. Security had not cleared us to be on the base so the only place we could get off was the Optics Building where we would be working soon. We met a few of the people working there, looked around, and within 15 minutes were back on our way to the base.

As we left, I felt both excited and disappointed. It was nothing like I had dreamed; but I was excited because I realized that hopefully within the next two weeks I would be at the cape, filming my very first launch. I was ready and looking forward to it.

Bill Weimer was a thin man of average height, no taller than 5'9". He sat next to me on the bus going back to the base and told me of his interest in rockets. We both shared the same enthusiasm about missiles and the cape, though Bill seemed a lot more knowledgeable. He knew all about missiles used during World War II and some of the newer ones . . . their names, sizes, and purpose. Bill wasn't married and had moved down to Miami from New York a year earlier to live with his mother and sister, then came to Cocoa when he was hired by RCA. He said he had read an ad in *The Miami Herald* for people to work at the cape and, because of his interest in missiles, drove up there one day just to see what was going on. A short time later, he began training as a camera operator/ technician, the same as me.

Bill looked to be about 30, four or five years older than me. He spoke with such excitement of how he loved to play the drums and build model rockets that I found it hard to believe he was even 18. He didn't have the manner or appearance of an adult. Physically, Bill looked his age; but his movements were clumsy and awkward. He reminded me of a grown-up kid, talking only of things that interested him.

Through the years Bill became a part of our family. He would frequently come over after work and on the weekends, sometimes just to watch TV and read the paper or at other times to show me a miniature rocket he had made. These weren't models; they were real missiles! Bill liked to experiment over at my house. I was always worried that one of those wild rockets would go flying into the neighbor's house and kill somebody!

For the next few days, I was back at the base, working in the optics shop in the mornings and tracking Air Force jets in the blinding afternoon sun. We were all getting tired of waiting . . . waiting for our chance to get out there to do what we had spent three months learning to do.

The waiting was affecting all of us, some more than others. Every day at lunch one of the guys with whom I was working kept talking about how some were taking all the jobs away from "us Southerners." George Hendricks was a cocky bastard to begin with, and this waiting took away what little patience he did have. You knew that if you ate lunch in the base cafeteria with George, you would have to listen to him complain about "those damn Yankees." He was from Georgia and apparently never got over what happened during the Civil War almost 100 years ago.

Most of the guys in the section originally came from up north many years earlier as I had, and we used to make fun of George and his country accent. For a while it was amusing for us to be called "damn Yankees."

"Ya'll should've stayed up north," he said to me one day. "We don't need 'neny' of you people down here."

"C'mon, George, knock it off . . . at least for today. I'm tired, and I don't feel like listening to it."

It was getting monotonous, hearing the same thing day after day. I didn't want to push George too much, but I knew he would keep it up unless somebody said something. Maybe that was the trouble: no one ever challenged him.

"Look, Yankee," he shot back. "I got the right to say anything I want. Ain't nobody gonna tell me to shut up . . . especially some damn Yankee!"

"George, give us a break," one of the other guys said.

"I didn't tell you to shut up," I added. "I was just asking if you could talk about something else. I'm getting tired of being called a Yankee all the time and hearing how we're all taking jobs away from the Southerners."

"What's wrong, Yankee? Don't you wanna hear the truth? M' brother can't find a job because of people like you. He was born and raised in the south, and he can't find no place to work because of all you northerners. Ya'll think you're smarter than us, don't ya?"

"George, once more, will you knock it off?"

I didn't want to start anything because there were majors, captains, and other Air Force personnel sitting all around us at the table.

"And what happens if I don't, Yankee? You gonna tell me to shut up or else? I'm gettin' sick of you fellas from the big city tell'n me what to do. Ya'll move down here and think you been here all your life . . . like ya own everything. I been in this part of the country a hell of a lot longer than you, pal . . . so don't tell me to shut up! And don't threaten me, either . . . Yankee!"

That was it! I couldn't take anymore.

"You son of a bitch!" I shouted across the table. "You call me a 'damn Yankee' once more, and I'll . . ."

"Sit down and shut up, Yankee!" George interrupted.

"You bastard! I've had enough of you and your southern shit! If you southerners weren't so damned stupid, us northerners wouldn't have to come down here and do your job for you!"

I was eating a banana at the time, and I threw it at him as hard as I could. As I stormed out, I could see him out of the corner of my eye, sitting there with this stupid look on his face and banana all over the front of his shirt. I almost started laughing, but I was too upset.

I never had any more problems with George. We saw each other several times again, before and after we were transferred to the cape, and got along fairly well . . . but he never pulled his "southerners routine" on me or anyone else again. He knew better.

Even though we received some harassment and overbearing comments from the local workers accusing us of coming down and taking their jobs, I found out later that our agitation paled in comparison to what others must have endured.

I was not fortunate to have personally known Mr. Julius Montgomery even though we both started working for RCA at the Cape Canaveral Air Force Station around the same time in 1956; however, I learned of his story: Mr. Montgomery was the first African-American hired in a professional position at the cape and a fellow Range Rat who worked on missiles and satellites, repairing the electronic failures. Mr. Montgomery would later speak of his trials and tribulations regarding the racism he encountered.

He described on his first day of work when he was introduced to his all-white co- workers, they would not acknowledge him nor shake his hand. His bouts of racism certainly didn't end there. He continued to experience casual and blatant racism in his work routine for years. If the racial demagogues thought their hatred and bias were going to detour Mr. Montgomery, they surely did not know the man.

In 1958 the Brevard Engineering College was a new institution, preparing to offer continuing education classes to the teams of engineers and technicians working at Cape Canaveral. As the new location and building had yet to be constructed, the classes were held in a public junior high school in the area. Once the Brevard County Superintendent became aware that Mr. Montgomery would be attending, the Superintendent demanded that Mr. Montgomery's attendance be denied. Otherwise, he would reject the college's use of the facilities.

Being the man he was, Mr. Montgomery withdrew his application so the college could open and, thereby, make it possible for his fellow team members to attend the school. He would have to wait until the college opened its own facility before he could attend. Mr. Montgomery didn't just break barriers and make it easier for other African-American men and women who came after him at the cape, but he went on to win a government seat with the City of Melbourne.

It wasn't long before we boarded a bus again for another look at the cape. Only a week had gone by since the last trip, but the change was incredible. Buildings seemed to have gone up overnight. There were hundreds of people walking and driving around all over. I saw several new roads, and more were being built. The land around the central part of the cape was still barren, but there was much more activity everywhere. It was beginning to take shape, looking more how I thought a missile base should look.

We had no idea why or where we were going this time. The bus had already passed the Optics Building and was heading back to the south gate toward Patrick Air Force Base. This was the same way we had come before. In a few minutes we would be back at the base. We found out the reason for the trip once we returned: As we filed off the bus, we were given a badge and an identification card with our name and security number.

"Okay, this is it," one of the instructors announced. "You're on your own. You know how to get there and back. Check the board inside to find out where you're to report and whom you are to see when you get there. You'll start work tomorrow. Make sure you have your badge and identification card with you at all times. You'll need to show it at the gate to get in and out and anytime a guard asks to see it. If you should ever lose your card, notify the security office immediately. You will not be permitted to enter the cape area without it. Good luck. If you have any questions, I'll try to help."

_____ Chapter 8

In less than 12 hours I would be getting ready to leave for my first day at the cape. Was I prepared? Those three long months of training suddenly seemed so short. Did I remember everything I learned? How much had I forgotten? Was there anything they forgot to tell us? All my apprehensions faded away as I thought of what it would be like to finally be working at the cape.

At 8 a.m. the next morning I pulled up at the Optics Building and met my leader, Earl Miller. Earl had been a pilot in the Air Force and flew in the Berlin Airlift before coming to the cape in 1954. He was a hometown boy raised on Merritt Island, graduated from Cocoa High School, and joined the service in 1939. He got in on all the cold wars: in Korea he flew B29s out of Okinawa and chalked up 148 flights in the Berlin conflict following World War II. Earl was initially hired as a chief theodolite operator and later promoted to leader of the section once it began to expand and RCA realized they had to have a supervisor.

In the years I worked with Earl, I never once saw him without his pipe. Lit or not, the pipe was always there. Earl moved slowly and talked softly—sometimes mumbling— and no one ever saw him get excited . . . even when he was almost killed by a missile that came back down one night.

The two years he had over me at the cape showed both in his age and the way he ran the section. Earl was as happy as the rest of the men in theodolites to see some new blood coming in. There was a huge turnover in personnel from 1955 to 1956, especially in theodolites. Guys were always quitting after one or two nights out in the field. I would later find out why.

There were about five of us new recruits who showed up the first day to the theodolite section. Half of the men I was in class with had been assigned to a different camera group as Bill Weimer and George Hendricks had, and there were three or four who never made it past final exams. Since all camera operators in the various sections worked out of the Optics Building, I saw Bill and the other guys quite often. We usually met when one of us was coming in and the other was leaving.

About 50 or 60 people were based in the one-story Optics Building. From this day forward we would report to this building for supplies such as film, camera parts, or electronics parts. The place was so crowded that if all 50 or so people went there at the same time, only a third would be able to get inside.

The documentary, ballistics cameras, motion picture tracking, CLR cameras, and theodolites sections each had a room off one side of the center hallway. An optics shop with four men was to the right of the back door at the end of the hall. They cleaned the cameras and handled minor repairs so we wouldn't have to drive all the way down to the lab at the base 15 miles south. During a launch, when time was extremely critical, it helped to have our own shop.

The theodolite section did not have as big a room as the other sections because we didn't need any space to store the cameras; they were all stationary in shacks out in the field. All we kept in the room were our long, gray, metal boxes which contained tools, film, and spare parts for the electronic consoles connected to the camera.

Earl took a few minutes that first day and introduced us to the other people in the section: R.J. Patskoski (Pat); Willie Cubbin, Earl's assistant; Barry Lough; Dean Waddell, the group manager; and Stan Purnell, the coordinator. The rest of the group hadn't shown up yet because they were out the night before on a launch.

Pat led us outside, and we hopped in the back of a three-quarter ton truck for a backroads tour of where we would be working. As we rode along, Earl sat in front with Pat, pointing out some of the buildings around the cape and explaining what they were used for. It was hard to hear over the roar of the engine and flapping canvas.

The truck was military blue like all the other jeeps and pickups, undoubtedly left over from the war. A khaki-colored canopy covered both the cab and the back. Twice I hit my head on it when the truck dropped into a hole, first the front wheels, then the back. I could feel vibrations from the left rear tire through the steel seat. Using trucks like this, I'm surprised we won the war: if the enemy didn't get you, the trucks certainly would.

Pat took us down a path worn through the heavy sawgrass by other trucks until we got to the beach.

"This is a short cut," Earl explained. "It's a lot quicker going this way than down the roads. You've got to watch out for the high tides, though. Otherwise, you're liable to get stuck."

It felt strange, sitting in the back of a truck as we drove down the beach . . . an odd way to get around a missile base. One thing for sure, it was a lot smoother than some of the so-called roads.

We were going to the nearest of the four theodolite sites, four miles south of the Optics Building near Port Canaveral. Using the so-called short cut, in less than 15 minutes we were there. After walking a short distance through the brush, we came upon a small wooden shed.

Earl and Pat showed us how to move back the green, wooden shed which covered the equipment in between launches. It was about the size of a storage shed, no more than 8′ by 8′ and set on two steel rails similar to railroad tracks. Pat explained that we needed to push the shed backwards to uncover the Cine theodolite tracking camera and the two electronic racks associated with it. We unlocked the two front wooden doors.

The doors were of the height of the 6′ shack and had to be opened before the shed could be pushed back behind the two electronic consoles. We referred to these as racks. The racks were huge, steel boxes which resembled stout, metal lockers. In the left box were the tubes and circuits; the panel on the right contained several meters and control switches. On top of the racks were a phone box and an oscilloscope. Two feet in front of the racks, bolted to the concrete slab floor, was a

Theodolite Camera

pedestal 3 1/2' high and about 2' in diameter. The theodolite camera was attached to a swivel on the pedestal so it could move freely from side to side. Operators, two for each site, stood on either side of the main lens in the center as they recorded the launch, peering through a scope.

One person followed the angle, or azimuth, of the missile while the other tracked the elevation. When filming a launch, each cameraman had to crank a wheel by the mount so the camera would move along with the object as it moved horizontally and vertically. The equipment looked new, as though it had just been painted. Everything was gray, including the cement slab. Despite being out in the open, the site was exceptionally clean.

After pushing the shack back to uncover the camera, I was given a quick orientation of the shed itself and the surrounding area. It was noted that on the left wall and the right wall there was a long bench that was to be used as our sleeping area during long launch holds. Each bench was about two feet wide and six feet long. Outside, Pat pointed to four target boards and the launch pad and gantry. Off on one side was a pile of wood for our burning when necessary, and during slack times we were to keep the wood pile well supplied and covered with palm fronds. There were no bathrooms in the brush when nature called and no way to get back to our building if a rocket went into a 1-, 2-, or 3-hour hold or longer.

I looked around as a few of the guys slid the shack back in place. How barren: there was nothing nearby in any direction except trees, scrub brush, palmetto thickets, and more trees. As they pushed the shed, my eye caught some animal or reptile scurrying out from underneath.

"Look out!" I shouted.

"Don't worry; I see it," Pat said calmly. "There're quite a few of them out here. They won't hurt you if you don't bother them."

What one of the instructors in class told us when describing the early days at the cape seemed to be true still today: Ralph Norris, an old-timer, always loved to talk about what it was like "back then." He had been there since 1949 and warned us many times about the varmints in the brush.

"Boys, when you get out there," Ralph would say in a gruff voice, "watch yourselves. It's not as bad as it used to be, but there's plenty of snakes out there. One time it was so bad that the Air Force made everyone carry a snake bite kit. The nearest hospital was at the base . . . a good hour away back then. We didn't have the roads we have now. A few guys almost died before the kits were issued. Fortunately, I never had to use one. Saw plenty of rattlers, though . . . too many. There are other animals out there, but I don't think you'll have to worry too much about anything else."

We covered areas we could walk into and areas not to be in. I also was informed as to what I was to bring to the site with me: a hat and sunglasses, along with water, rubber bands, and toilet paper topped the list. It wasn't long before I found out the rubber bands were to be used around my wrists and ankles to keep mosquitoes out of my clothing.

Pat informed me that after arriving at this tracking site, I could be here for a long time before returning to the Optics Building. This was due to the cape's scheduling a launch, for example, at 9 a.m. and another launch at 12 p.m. and yet another launch at 3 p.m. If the 9 a.m. launch went into a hold and continued longer than 12 p.m., then we set up for the 12 p.m. launch. If that launch went into a hold until 4 p.m., we then had three launches at different launch pads on hold at the same time. It could be 4 p.m., and we still hadn't launched the first scheduled launch at 9 a.m. This could go on all through the night and into the next afternoon. I remember spending many hours at the site due to long holds and new launches being scheduled.

With this in mind, we needed a program for survival regarding food, water, and appropriate clothing. If we were caught at the site for exceptionally long periods of time, the truck would make a run for some type of food.

The cape during that time was a jungle. Nighttime was "alert" time. To be able to see around our area, we would build small fires. We never strayed far from our shed. If the launch hold was long, we would pull the shed over the camera equipment and both doors would be closed while we took turns sleeping. There were wild animals all around us: rattlesnakes and other varieties of snakes, bobcats, skunks, and the Florida panther were our main concerns. Many times driving to our site at 1 or 2 a.m., a Florida panther jumped across the narrow road in front of our truck.

That was our field training. I considered it a survival course. All and all, however, my first day at Cape Canaveral was exciting and, of course, very informative as to what my future looked like.

 Chapter 9

When we returned to the Optics Building, we were informed that a Redstone rocket launch was scheduled for late afternoon of the very next day. Earl paired us off with the other guys in the section. I would be working with Pat, one of the four site chiefs.

Pat impressed me even though I met him only an hour earlier. He seemed like an easy person with whom to get along . . . always smiling, always happy. After working with him a few months, I felt I had known him for years. Each time Pat came to work, he would be wearing the same, worn, baseball cap and carrying his lunch in a brown paper bag. Many times, I wanted to ask him if he slept with his hat on.

Everything to Pat was okay. If anything ever went wrong, he would tell me, "Don't worry about it; it'll be okay. It's all taken care of; it's okay."

I guess he figured he was in charge of the site and would take care of the problems.

Pat took his job very seriously; everything had to be done just right. We spent many hours onsite, waiting for various launches, trading stories about what we had done before coming to the cape and talking about the many things we had in common: we enjoyed fishing; we were married within four months of each other; we both had a baby daughter; and we were both the same age. Pat was in the Air Force before joining RCA. He and his wife were living in St. Petersburg and moved to Merritt Island in August of 1955 when he started working at the cape.

Before we did anything else, Pat showed me a map on the wall in the theodolite room. He pointed out where each of the four camera sites were in relation to the launch pads and other buildings around the cape.

The black-and-white sketch merely outlined the Cape Canaveral area from Patrick Air Force Base to the northern border by Titusville. At each end of the cape, 13 miles apart, the land narrowed down to less than a mile wide between the Atlantic Ocean and the Banana River on the west. The widest point of the cape was in the center and looked to be about five miles wide, jutting out into the ocean.

Most of the buildings and roads shown in the sketch were in the industrial area near the Banana River, halfway between the north-south boundaries. One of the theodolite sites, 1.4, was located there in the industrial complex on top of the central control building. The other sites were 1.3 at the north gate of Patrick Air Force Base; 1.2, Earl Miller's site, at the southern tip of the cape near Port Canaveral; and 1.1, Pat's site, about five miles north of the Optics Building.

When Pat first drove me to his site, I was amazed how far away from everything it was. Going through five miles of near wilderness made it seem farther. There was nothing further north except old, abandoned houses which the Air Force bought from the residents many years ago. Between launches, sometimes two or three weeks apart when there wasn't much else to do, we would rummage through the dilapidated buildings, taking what little there was left inside.

From where we were, I could barely see the lighthouse which was near the coast in the center part of the cape. I knew that the Optics Building, where we kept our supplies, was just to the west of the lighthouse. We were out in the middle of nowhere! Whoever decided to put a site here had to be crazy. At least the other three sites were near something: the Air Force base, Port Canaveral, or central control. But there was nothing here . . . not even a paved road!

The ocean was to the east, 100 yards away. Every place I looked there were clumps of scrub brush, sawgrass, and trees. What a deserted place! I wished I was back in class. Better yet, I wished I was back in that secure, air-conditioned office in data reduction back at the base. I didn't even want to think what the site would be like at night. No way would I ever go out there alone. Perhaps it wouldn't be that bad out there with somebody like Pat.

It wasn't too long before a launch was scheduled. The Air Force was planning to test a modified Redstone rocket which was one of the first ballistic missiles developed by the Army in 1953. The rocket got its name from the Redstone Arsenal in Alabama where it was constructed. Some of the early Redstones were brought by truck from the arsenal to the cape, and it was not unusual to see pictures in the local papers of a convoy traveling through small towns along the way. Back in the early 50s, the scientists who designed the missiles lived near the Redstone Arsenal and came down to the cape only for launches. When suitable facilities were built later, the launch team moved its operations to the cape area; and the Redstones were assembled in Hangar C.

This would be the first launch I would be actually tracking. A few weeks earlier, right after graduating from more than 300 hours of training, I was at a ballistics camera site in Canova Beach for launching of the first Jupiter C, an Army-Navy variation of the Redstone. All I did that night was help set up the site and load equipment on and off the truck. Although I didn't know it then, Jupiter C set a new first: it was the fastest (15,000 miles an hour), farthest (3,300 miles), and highest (650 miles) of any man-made missile ever launched. The missile took off shortly after dark and plunged into the Atlantic close to South America.

The missile I would be tracking would probably be a modified Redstone, similar to the Jupiter C. We spent all day Monday making sure we were ready for the launch.

The next day I arrived at the Optics Building three hours prior to launch time. The weather was perfect. We were ready!

No one had been out to the 1.1 site in the last four days, so Pat had suggested we take fresh chemicals rather than taking a chance on using the developer and fixer that had been left there from the last launch. Before each test we would have to run off 10 or 15 frames of film and develop it onsite. If the chemicals were old, the film wouldn't develop properly, and we wouldn't be able to check the exposure on the camera.

By the time I mixed the chemicals and put them in gallon bottles, Pat already had finished rolling the 35mm black-and-white film into the gray metal cartridges. We usually took a case of film, eight rolls of 100' each. Although we only used one roll on a launch, we'd go through three or four if something went wrong with the racks and we had to keep running tests. In addition, there was always the chance—which happened frequently—the film might get jammed when we loaded the camera; and it would come flying out all over the site.

Pat watched me as I checked if all the tools were in the small kit inside the metal storage box. If any of the tools were missing, it would be almost impossible to fix any part of the theodolite. The kits had been sent over from Germany, along with the cameras; so everything was in the metric system. I had a hell of a time working on the cameras whenever something went wrong. All my life I was used to 3/8", 7/16", 1/2" 9/16", 5/8", 11/16", etc. Now I had to work with wrench sizes like 9mm, 10mm, 11mm, 12mm, and so on. After a while it was easy; however, in the beginning when it was more important, I would always get the sizes confused.

Pat put the film in the box next to the kit. We signed out on the log and went outside. Most of the other guys were already in the truck waiting.

For a long time, there was only one truck for the entire section. Earl would drive everyone around to the four sites and then take the truck down to Port Canaveral where he usually worked. After the launches, he came around again and picked us all up.

We left the Optics Building about 1:30 p.m. and headed for site 1.4 on the second deck of the central control building where all launch systems were monitored, including the theodolites. Two guys got off there, along with Willie Cubbin, Earl's assistant, who would be checking with each site that afternoon to make sure all cameras were working. Willie would turn all the cameras on simultaneously at lift off.

When Earl wasn't at his site at Port Canaveral, he would be at 1.4, the master site, controlling the cameras. Instead, for today's launch Earl was going back to the Optics Building after dropping us off. There were enough people to cover all the sites, and Earl didn't have to go out. If there were any major breakdowns in the equipment, Earl would be able to bring out replacement parts or fix the problem himself. The Optics Building was more centrally located among the four sites than the master site at central control, and each of the sites could be reached within half an hour.

As we neared Pat's site, 1.1, it was about 2:15 p.m. Earl still had plenty of time to make it to the two other sites to the south and get back to the Optics Building before the 4:00 p.m. launch.

The truck headed off the pavement onto a dirt road toward the shack. Pat reached into a large paper bag and pulled out a Mexican sombrero with cheese cloth hanging down around the sides and a package of rubber bands. What the heck was he doing? A few of the other new guys in the back of the truck started to laugh. I think they were amused at the look on my face, rather than at what Pat was doing. Before I realized it, Pat wrapped rubber bands around my pants at the bottom and on the cuffs of my long-sleeved shirt.

It was only late October; and even though the nights were still very hot, Pat had told me to wear a long-sleeved shirt because of the mosquitoes. We had been out to the site a few times in the past month, and the mosquitoes didn't really seem that bad. As he was tying himself up in the same manner, Pat explained it had rained the night before and there would probably be a lot of mosquitoes out because of the swamps all around the site.

Suddenly the truck stopped. Pat grabbed a portable tank sprayer and leaped out the back.

"Follow me," he shouted as he ran toward the small shack. I reached for our toolbox and a few other supplies and tried to catch up to him. I could barely see Pat through the foggy mist of insect repellent and thousands of mosquitoes flying up from the thick sawgrass. It was incredible. Never had I seen so many mosquitoes in one place! Swarms of them buzzed around my head and landed on my face. Breathing became hard; I had to use my nose, praying I wouldn't inhale any of the bugs, and remember to keep my mouth shut as tightly as I could. Dozens of mosquitoes gathered around my eyes, but there was nothing I could do. My tan shirt and brown pants were almost black from the legion that attached itself to me. I could understand why Pat tied my pants and shirt cuffs. Without it I would have been even more vulnerable. The thought of these things inside my clothes, gnawing at my skin, triggered a subconscious panic inside me . . . like I was living some sort of nightmare.

Pat put down the sprayer next to the front doors of the green shack to help me with the toolbox and chemicals I was carrying. When I reached the shed, Pat told me to open the two plywood doors while he continued spraying. I set the rest of the supplies down and quickly began trying to brush off some of the mosquitoes. Most were already dead; the few that were left flew off. Even after Pat again covered the entire area including the shed, still the air was filled with mosquitoes— though only about half as bad as a few minutes ago.

 *Chapter* 10

"Looks like we might have some problems with the shack," Pat told Earl. "Hold on a minute."

We opened the doors and began pushing. Nothing; it didn't move an inch. Sand had built up in the back by the rails probably from the rain we had the night before.

"We need a push!" Pat shouted.

"Okay. Stand back," Earl answered as he backed up the truck and pulled around in front.

Slowly, he eased the front bumper closer until he hit the front of the shack. Earl threw the truck in second gear and popped the shack back about a foot.

"That should do it," Pat responded. "Thanks. Talk to you later." After a brief pause, Pat explained, "Sometimes we have a few problems with the tracks."

We shoved the shack back the rest of the way.

"What happens if the truck isn't around?" I inquired.

"They usually wait to make sure we get them pushed back before they leave. Otherwise, they have to come all the way back out."

Pat started to load the camera.

I said "Hey, let me give you a hand with that."

"That's okay; I've got it," he responded.

"Anything you want me to do?" Since I had never been out on an actual launch, I wasn't exactly sure what to do next. I felt guilty standing there while Pat did all the work.

"Don't worry; you'll have plenty to do later. Besides, I'll be finished in a few more minutes. By the way, you said you went through training, didn't you?"

"Sure did . . . three long months. I thought they'd never get around to putting us out here!"

"Three months? What did you do all that time?"

"I think we learned everything there is to know about this camera. Maybe they thought we were going to build the cameras instead of just film the launches. I hope I never have to go through that again. I haven't studied that much since I was in high school . . . four and five hours a night including weekends!"

"Did they let you work the cameras at all?"

"Yeah. For about the last month of training, we went over to the beach across from Patrick Air Force Base every afternoon and filmed jets taking off and landing. I used to bake out there, sitting in that hot sun . . . especially in the middle of the day."

"I can imagine. It gets pretty bad here in the summer. There, that should do it; everything's hooked up. Want to turn the power on?"

"Sure!" I quickly said.

Then I remembered, I didn't know how.

"Uh, where's the switch?"

"Oh, sorry. It's over there. See that box on top of the post right next to the palmetto stump?"

"Right. Be back in a second."

I was just about to open the top of the wooden box when Pat ran over.

"Stop! Don't open it! There's something I forgot to warn you about."

Immediately, I thought I'd done something wrong.

"Any time you go to turn the power on, throw the cover off and jump back. It's warm in there, and lots of times rattlesnakes like to curl up inside."

"You mean there are *really* rattlesnakes out here?"

"Sure, they're all over in these weeds."

"I thought everybody was kidding me!"

"Believe me, they're real. I've seen more than my share of them. I've never had any problems . . . if you don't bother them, they won't bother you."

We walked back to the site and began setting up. The launch was in an hour and a half. In less than 15 minutes everything was ready. Pat watched the meters on the racks as I moved the camera around. It was working fine . . . no problems.

"All right, let's do a test," Pat said.

He went to his side of the camera which tracked the angle. Aiming at one of the poles we used to line up all the cameras, Pat turned on the camera and let it run for three or four seconds. We needed only about 10 to 15 frames to check our film, and the camera shot four frames per second.

I pulled the film cartridge out of the camera and put it in the dark bag. It was daytime, and I would have to be extremely careful loading the film into the developing tank. Once it was in the can, I could add the chemicals: first pour in the developer; throw that out; wash the film with water; and then pour in fixer so the film wouldn't fade in the light. The entire process took about 10 minutes.

Pat scanned the negatives after they were dry. There were several things he had to check: exposure, timing, elevation, and the angle all had to be recorded in each frame. By reading the timing, which showed up as a series of long and short black marks on the edge of the film, one could tell exactly when the frame was taken during the launch. The elevation and angle also had to be on each frame. Tiny lamps flashed as the film passed through; and small numbers, showing elevation and angle, would be recorded in each lower corner of the 35 mm frame.

If everything worked right and the object upon which we lined up was centered, we then waited for pre-launch orientation. We couldn't go into orientation until 30 minutes before a launch to make sure the equipment was still level and lined up.

I could barely see the missile from the site. I hadn't viewed it through the magnified scope yet because the camera was set on one of the target boards, waiting for orientation. The Redstone would be launched around Complexes 5/6, halfway between the Optics Building and the theodolite site at Port Canaveral, 1.2.

"How far do you think we are?" I asked Pat.

"Oh, about seven, eight miles."

"I didn't think it was that far. Do these things usually go off when they're supposed to?"

"It depends. Sometimes there's a hold, and we must wait until they get whatever's wrong fixed. It's not too bad . . . once you get used to it."

"Any ever explode?"

"A few. I've seen about maybe two dozen launches, and I 'd say about five or six blew up."

"Anybody get hurt?"

"Nope . . . not yet. They're pretty careful about letting people get too close to the pad. I think there's a mile . . . mile-and-a-half . . . safety zone around the pad; and no one can go in or out after a certain time."

I thought about what it must be like inside the blockhouse, just a few hundred feet away from the missile. About a week earlier, I met one of the technicians who worked in the blockhouse during the launches. He said there wasn't too much to worry about: the walls were 2′ thick, and the ceiling was 8′ thick. The narrow windows facing the pad were made of three, 4″ panes of bulletproof glass. I couldn't realize what it would be like locked up inside there with 70 other people for two hours, waiting for a launch. One thing, though: they had protection . . . we didn't.

Suddenly, Willie Cubbin, Earl's assistant back at the control center came in over our headsets, "All sites prepare for orientation."

Using the headsets, we were automatically connected to everyone else at each of the other theodolite sites.

"Okay, Willie."

"Ten-four, Willie."

"Roger that, Willie."

I could hear as everyone checked in.

During orientation all the cameras would line up at the same time on each of the four or five target boards and shoot a few frames. Willie turned the theodolites on and off at the master site so each camera would have the identical number of frames on each board.

The boards were similar to telephone poles, about 5′ high, with a piece of plywood nailed to the top. When film readers in data reduction got the film from all the sites, they would compare the angle and elevation on each roll, giving them a three-dimensional computation of the missile's flight. They would use our orientation shots of the target boards to begin their calculations. The bobcats in our area liked to be on the top of our target boards; so, while aligning our cameras, we would also be taking many pictures of bobcats.

We finished shooting the target boards and lined up on the missile.

"All sites operational," Willie reported. "Standby for countdown. Launch in minus 19 minutes."

The missile looked enormous through my scope! I could hardly see it before, and now it filled the entire sight as though it was just a half mile away. It was magnificent! It was just as white and slender as I pictured.

"What are the black stripes on the bottom part for?" I asked Pat without taking my head from the scope.

"For tracking . . . so they can tell if it's spinning or not. Most of the birds have them."

I studied it: the first missile I would be tracking. Liquid oxygen poured out of an opening midway up the side. The red gantry was still in place. A minute or two before the launch the gantry would move back. An umbilical cord, which fed the oxygen into the missile, would remain attached until seconds before liftoff. Then it would drop, and the opening below would be closed. Another chemical would mix with the liquid oxygen and ignite. I had learned all this while I was in training.

"T minus 15 minutes. All systems go," Earl came over the radio.

I became more excited as the countdown narrowed. Admittedly I was scared. My fear suddenly overcame the ambition I had to work around missiles. The dream I held since childhood when I used to watch Flash Gordon each week at the movies was just that: a dream . . . where no one ever got hurt no matter what happened. But this was reality.

"Hey, Ed, you okay?" Pat asked. "You don't look so good."

"Oh, yeah, I'm fine. A little nervous I guess."

"I know how you feel. I'm the same way every time I see a launch. It's something you never get over no matter how many you see. Don't worry . . . everything'll be okay. We're pretty safe out here."

Damn! Why did I ever want to do this? I wanted to turn and run, but my legs wouldn't move. I couldn't let go of the camera. All I could do was stand and stare through the scope at the missile.

"T minus 14 minutes and holding. We have a hold."

"Ah, come on, Earl," one of the guys commented over the headset. "What the hell is going on up there?"

"Got me. Someone just called in from the blockhouse. Said there was a problem with something or other; I don't know what the heck's going on out there."

"Any estimate on hold time?" Pat asked.

"Yep, just got it. It's now scheduled for 5:30."

"What? Five-thirty? What are we supposed to do until then?"

"How am I supposed to know? They say it'll take that long to fix whatever's wrong."

There was nothing Pat and I or any of the other guys in the section could do. There wasn't enough time for Earl to come around and pick us up and then go back to the Optics Building. All we could do was sit around and wait . . . talk and wait. Once in a while I would get up from the wooden bench inside the shack and move the camera around . . . first, over to one of the boards; then, to the lighthouse near the Optics Building; and finally, back to the missile. Pat suggested I try my hand at tracking some of the pelicans along the coast. I swung the camera toward the ocean and picked up one bird just coming up from the water. It was easy to track the pelicans when they flew above the water. However, once they started to dive to pick up a fish, it was almost impossible to move the camera that fast.

At 5 p.m. we ran through orientation again.

"Okay. We're on countdown once again," Earl said. "Picking it up at T minus 14 minutes. All sites please report status."

"1.1 operational."

"1.2 operational."

"1.3 operational."

"1.4 operational."

"Roger that. All sites operational and ready. Standby. Going on T minus 13 minutes and counting."

Thirteen minutes away! I felt excited, nervous, prepared . . . and more than anything else, scared. It was lonely out here. Looking through the scope, all I could think of was two things: me and that missile. I felt I was totally alone with that thing. Nothing else in the world . . .

"Hold it! We have another hold! Holding at T minus 12 minutes," Earl reported. I could hear everyone groaning.

"What's going on up there?" someone yelled over the headset. "Don't they know what they're doing?"

"Same problem as before. Something to do with the lock for the liquid oxygen. It won't close for some reason. Might be frozen. New launch time is 8:05 p.m. Have fun out there."

Chapter 11

Earl knew what it was like out there in the field. He had been there many times himself. As soon as it got dark in another hour or two, the mosquitoes would come out in full force. They weren't too bad during the day, but at night you were at their mercy.

"Thanks a lot, pal," Pat said sarcastically to Willie in response to his new "hold" radio message.

"Hey, Ed, better go back into the shack and get the sprayer. I'll spray again; maybe it'll help a little."

"They get that bad?"

"Just wait and see. I had a guy working with me last year who forgot to bring the sprayer. I had a can of insect repellent, but it didn't help at all. He had on white socks; they turned bright red from scratching all the mosquito bites He stayed that night but quit the next day."

Pat sprayed the ground around the site, and then we went out looking for palmetto stumps while there was still enough light to see. Many of the early settlers in Florida found that smoke from burning palmetto stumps helped keep away some of the mosquitoes. That was about all one could do to keep the mosquitoes away. Nothing could kill them . . . not even the fogging trucks which sprayed the cape area twice a week.

Numerous times back in 1949 and 1950, all activities and construction were stopped because the mosquitoes were so bad. One of the other theodolite operators, Barry Lough, who lived in Florida most of his life, said the mosquitoes and sand fleas were so heavy at a Navy site in nearby Fort Pierce during World War II that several men on night watch committed suicide. Even now, the Air Force issued sprayers to all employees at the cape, and the county provided free repellent to everyone who wanted it. I was starting to understand why my neighbor from Boston was being paid "swamp pay."

For the next two hours, Pat and I just sat in the shack, talking. Every once in a while one of us got up to spray. I was getting eaten alive, and the smoke from the palmetto stumps was coming right at me. The smell was nauseating. The only thing it seemed to do was make me sick. It didn't seem to bother the mosquitoes at all.

Fortunately, Pat had brought a lunch with enough food for both of us. I didn't bring anything to eat. The launch was scheduled for 4:00 p.m., and I planned to be home by 6:00—maybe 7:00 at the latest—in time to eat dinner. I was so hungry by 7:00 that the bologna sandwich Pat shared with me tasted like steak.

It was going on 7:30 p.m., almost dark. The site felt more isolated than ever before. The darkness cut us off from even the trees and shrubs just 10 feet away. The dim light hanging from the racks barely lit our site.

We went through our third orientation. All sites were operational . . . except ours.

"What's wrong?" Earl asked.

"I don't know," I replied. "Pat's working on the racks now."

"You've got about 20 minutes yet. Can you have it fixed by then?"

"How the hell should I know?" Pat yelled into the headset. "I don't even know what's wrong. As soon as I find out, I'll let you know. We're not getting any pulse at all!"

The camera didn't go off and on when Willie flipped the switch during orientation. We could track the missile, but we wouldn't have any pictures of it. Pat searched all over the racks, testing the connections and checking the meters. Everything seemed to be okay. I could tell Pat was getting nervous.

"Two orientations and the damn thing works fine. Now we're ready for a launch, and it doesn't work," Pat mumbled.

"Find out what's wrong yet?" Earl came back.

"No! I told you I'd call when I did."

"All sites, get ready to pick up the count," Earl announced.

"Picking up the count at T minus 12 minutes. All sites operational except 1.1"

Earl turned off the headsets and called us on the private phone on top of the racks. Whenever the phone rang, we had no choice but to answer it. The bell continued to ring until someone picked up.

Pat yanked the phone down. He knew it was Willie.

"What do you want? I'm trying to fix this thing!"

"Do you know what's wrong yet?"

"Well, it's too late to send Earl out; there's not enough time. You're on your own. I'll check with you in a few more minutes."

Then I heard Willie come over the headset again, "Counting, T minus 10 minutes. All sites operational except 1.1. 1.1 is down."

Pat pulled off his headset and threw it on the ground. Suddenly, the phone started ringing again.

"What now?"

"Did you find out what the problem is? Is it working yet? Can you get it fixed?"

"Damn it, Earl!" Pat screamed. "I'll call in when I get it fixed!"

He slammed down the phone and went back around the racks, trying to see if anything had come unplugged.

"Pat," I started, "the count just came over. T minus six minutes."

"Okay. I don't know if I can get this thing going in time. I'm gonna keep working on it. You just track in case I get it running . . . and don't lose it!"

"All right."

I was ready. I kept my eyes on the missile. The liquid oxygen coming out of the side was turning to ice. The spotlights shining up from the ground reflected off the frosty, white metal. Since Pat couldn't work his side of the camera, I had to reach around with my other arm and release the wheel which followed the angle of the missile. I kept hoping Pat would get the racks working. My first launch and no film.

"The count! T minus two minutes, Pat!"

He didn't seem to hear me. The gantry slowly moved away from the missile. I could hear the loudspeakers off in the distance warning everyone in the immediate area to take cover. The shrill of the horns shot through me like a piercing scream.

Faintly I could hear the announcement, "Clear the area! Clear the area! This is your last warning. All personnel, please clear the area and take immediate cover."

It was an eerie sound echoing through the empty night.

The phone rang again.

"How's it going? Get it fixed yet?" Earl asked Pat.

"No! How the hell can I get anything done if you keep bothering me?" Pat placed the receiver on top of the rack next to the phone.

"He won't bother us anymore," he remarked with a smirk on his face.

Over the headset I heard all the sites report in for the final countdown. One minute left. Pat was still working on the racks.

"All sites operational except 1.1," Earl announced. "Prepare for launch. T minus 20 seconds and counting."

"Pat! It's going! Twenty seconds!" I screamed even though Pat was only two feet away.

My eyes started burning from looking through the scope. I didn't want to blink; I might lose it.

"Make sure you stay on it," Pat ordered. "Don't lose it!"

I repeated the countdown to Pat as it came over the headset, "Ten! Nine! Eight! Seven! Six! Five! Four! Three! Two!"

The hose pumping in the liquid oxygen dropped, and the release valve on the side closed.

"One! It's going! It's going!"

"God damn it!"

Pat jumped up and kicked the racks. Glass from the meter covers flew everywhere.

"Damn thing! Stay on it! Don't lose it!"

"I got it! I got it! Don't worry; I got it! Look at that mother! I'm on it! I had it in the center of my scope until it went into the clouds."

"What's going on?"

"Pat, it disappeared! It's gone!"

"Keep the camera moving! Don't stop!" Pat ordered.

"There it is! I got it back again. Hey, what the . . . my God!" I yelled. "Pat! Pat! The flame's coming out the top! Pat, what's going on? What's happening?"

"What?" Pat yelled.

"The flame' s coming out the top! Pat! What's going on?" I repeated, urgently seeking an answer that would make sense.

Pat jumped over to the other side of the camera and looked through the scope.

"It's coming down!" he yelled.

"It's what? Where? Where's it coming down?" I pressed with panic in my voice.

I had no idea how far away the missile was by looking through the scope. I kept taking my head out, trying to see exactly where it was going to land. It was coming right at us! It was supposed to go the other way . . . downrange!

"Put your head back in there! Don't take it out again. Damn! That thing better hit soon. Keep tracking!"

"Pat! It's coming right at us! What'll we do? We're gonna get hit!" I screamed.

"Don't lose it!" Pat ordered again.

We heard a short "whoof," and we dropped to the hard concrete floor the same time the missile crashed. Burning scraps of metal flew all through the dark sky. Pieces rolled on the ground, bumping quickly toward us because of the trajectory of the missile. I heard pieces of the rocket land all around. A large heap of crumpled metal in flames lay on the ground, seeming to be less than a mile away. I was still shaking when we got up off the slab.

"Oh, no," Pat said in disbelief.

"What's wrong?" I pressed.

"Look at where it hit! That's right by the Optics Building! Earl and the other guys were all there! Pat grabbed the phone off the top of the racks.

"Earl! Earl! Are you there?" Pat screamed into the receiver.

No answer. There was nothing coming over the headsets or the phone. What had happened? Did it hit the Optics Building? Did Earl and the other guys make it out? After about five minutes the phone rang. Pat grabbed it.

"What happened, Earl? Anybody hurt?"

"Don't know yet. Too soon to tell. It landed near the lighthouse and the Optics Building. It's quite a bit west of Hangar C. Earl's okay. He said he'd be out to pick you up soon."

"Thanks. See you later."

Neither of us said very much as we closed down the site and pushed the shack back in place. We were both thinking. That thing landed too close. What would ever stop one of those things from hitting a site? I wanted to know why they didn't destruct it in the air, but somehow that didn't seem that important right now. We could have been killed.

Very few of the guys said anything about the crash as we headed back to the Optics Building in the truck. Most of us were new, and this was our first launch. Suddenly we realized what we were getting into and what the job was all about. I was scared. I wanted to quit. I loved the job and the chance to see the missiles . . . but not if it would cost me my life.

Luckily, I wasn't a mile or two farther away. The shock waves had gone over us because we were so close. I later found out the blast from the crash shattered windows in some of the nearby houses and was heard 60 miles up and down the coast and 25 miles inland!

When we returned to the Optics Building, we were asked to report to the section manager and his assistant for a debriefing. During our talk, I was asked if I was okay. He then grilled Pat about my performance from beginning to end. Before our leaving the section manager's office, he asked if I would be back in the morning because two launches were scheduled for the day and I was needed onsite.

I answered, "For sure!" and we shook hands.

This lasted for the next 14 years. I have no idea how many men resigned that evening. Pat simply answered my question with "Just a few."

I signed out on the log and turned in our test film. All the way home I relived the rocket crash. I kept seeing hot, burning metal flying through the air and rolling toward us on the ground. I hated the thought of burning to death. That had to be the worst way to die.

Security was extremely tight; everything that happened at the cape was top secret. I couldn't tell Jan what went on that night even though she would be able to see something was wrong. What kind of a life would we be living? Maybe it was better if she didn't know what happened.

I was only home a few minutes when she told me that all the neighbors came over that night shortly after the launch. Most of the people were wives whose husbands worked at the cape. She thought it was strange why they all came over so late and left right after I called to tell her I was on my way home.

I realized that one of the guys thought the missile had hit our site and called his wife. They didn't want Jan to be alone when she heard about the crash. Everyone thought I had been killed! Jan didn't even ask what was wrong when I got home; she didn't have to . . . she already knew. It wasn't until I went in the next day that I found out *all* the things that *really* happened that night.

_____ Chapter 12

That was the night I was introduced to the "Range Rat Society." I was now a member of a team of men and women who worked out in the open with no cover and roamed the cape like rats.

The unrecognized-hero Range Rats were the most incredible, dedicated, unselfish, and hard-working group of people who sacrificed themselves and time with family while living and working in primitive surroundings to advance America's mission of space exploration. Range Rats were as important as von Braun, the astronauts, the scientists, or other key players in the beginning of the space exploration era. They obtained key data from the rockets and space vehicles which was sent for analysis to Patrick Air Force Base and utilized to improve the performance of the next generation of space vehicles.

In the beginning, missile testing was typically short-range; however, as missile development continued, so did the distance they would travel. Thus, the development of the Atlantic Missile Range, which was later renamed the Eastern Test Range, began. The tracking stations and Range Rats were deployed from the cape to the Grand Bahama Island; Eleuthera; San Salvador; Mayaguana; Grand Turk; Dominican Republic; Mayaguez, Puerto Rico; Antigua; East Island, Puerto Rico; St. Lucia; Fernando; and Ascension Island near Africa . . . a range of approximately 5,000 miles.

Although these island outposts were officially called Auxiliary Air Force Stations, they were just referred to as tracking stations by the workers, which was the primary purpose. Early on, living conditions on some of these remote islands was extremely primitive as they lived in tents and Quonset huts which were built using galvanized steel until more permanent buildings could be constructed. As you can probably guess, there was typically no air conditioning or insulation; and inside temperatures could be sweltering in the heat and described as living inside a drum when it rained. Most of the workers may have appeared to be a motley group, dressed in shorts and t-shirts; but they still took their jobs and expectations seriously.

Working on the range was considered more than just a job by most of the Range Rats. It was a sense of duty and dedication to America's space exploration race. All the Range Rats felt extremely proud of what we did and the risks we took. It left all of us with a sense of purpose and the feeling that we were contributing to America's goal. There was a part of these men and women in every launch vehicle and spacecraft that left or attempted to leave Cape Canaveral. I was certainly privileged and proud to be considered one of them. Buzz Aldrin, Gemini XII and Apollo XI, said, "I salute from my space helmet the Range Rats, the unsung heroes of the space race to the future."

Throughout my career there were many formal certificates, awards, and recognitions of our work. It's funny how sometimes it is the non-formal recognition that means the most. Being known as a "Range Rat" was absolutely one of the things that made me most proud. Even though I had other official work titles, one of my best memories and greatest moments was when I received my own certificate that I was an official Range Rat. I cherished it then, and it is still framed on my wall.

A July 2005 article by *The Gainesville Sun* writer Gary Kirkland described the Range Rats well:

> *"They were seldom the smiling faces in the spotlight, but when the United States and Russia were running neck-and-neck in the space race, the Range Rats were hard at work, trying to give America an edge. 'Range Rats' was the nickname for a rough team of workers who tracked the missiles and rockets fired from Cape Canaveral. Working*

Range Rat certificate

from tiny island outposts across the Caribbean and Atlantic almost to the coast of Africa, they were a vital part of America's space-race team. The main task for the Range Rats was to track the missiles in flight and receive data from the onboard sensors and relay it back to Cape Canaveral."

Many a day I went to work early in the morning and didn't get home until late at night or even days later. My days many times included adventure trips downrange from Cape Canaveral to Ascension Island 5,000 miles from the cape.

When they placed the birds' (missiles or rockets) countdown on hold, we just had to hunker down and wait it out, no matter how long it took. We lived among the wild animals which included the Florida panther, rattlers, and bobcats. We were the Rats out in the open . . . no permanent buildings, yet; no food for long periods of time; and no toilets or toilet paper. (Thank God for leaves and palm fronds!)

I spent every day or night out in the open with rocket launches and many failed launches and explosions, day after

First launch blockhouse at Cape Canaveral

day, with the hope that one would not get us. In those days, not many rockets worked well . . . some of them blew up overhead, and the strong winds off the ocean blew pieces back over us. It was always a problem. The Range Rats were always in harm's way with no place to run.

Some nights we would have one hell of a rocket explosion; and the next morning the newspaper and radio would have a lot of commentary about it, not knowing anything at all. They reported that everyone was fine and that no one was hurt because they were in special blockhouses with 10-20' thick walls . . . well, not so much. There was an original blockhouse near the launch pad, but most of the Range Rats tracking the rockets were in the middle of the palmetto scrubs with no protection while sweating it out.

Not everyone was cut out to be a Range Rat. One particular night I had a new man working with me who had never been on the cape until that night. A Jupiter rocket was launched, and the programming of the rocket failed. It headed right back down toward us with the engine still burning. It would not destruct.

I can still hear over our com net, "Field personnel . . . take cover. No destruct!"

It came in right on the beach. It was a pretty good distance from us, but it sure looked like it was coming in on us. The new man was out of his mind. He started crying and yelling for me to do something and asking why I was not taking cover here or there. Of course, I acted cool, professional, and brave like any seasoned, veteran Range Rat (as Pat had done with me); and I didn't tell him the same thing happened to me on my first launch and left me shaking in my boots, thinking I was going to die. I told him it was useless and wouldn't do any good for protection. He ran anyway. I told him to hit the ground and hug it which is what I was doing. When we got back to the photo headquarters, HE QUIT and was escorted off the cape.

Most of the original Range Rats from the 1950s and 1960s unfortunately have either passed away, are in extremely poor health, or elderly like me at 92. For many years many of the Range Rats held reunions and established a social media site, sharing their amazing stories.

I was fortunate to meet up with Bill Weimer still living in Cocoa Beach, but I should have realized he never would leave the area . . . not even if he lost his job at the cape. Where else could he go to be around missiles and rockets? Bill always reminded me of a kid who never grew up. There were only three things he was really interested in: watching the launches; building model rockets which actually worked; and playing the drums.

It didn't surprise me that the first things he talked about were his model rocket collection and a launch he saw a week earlier. The only change in Bill was his appearance: he was older physically. He now worked at a marina, repairing boats. In his mind, though, I don't think he ever left his job as a film technician. He still went to the beach to watch every launch and knew all about several new missiles being tested by the Air Force and NASA.

Bill gained a lot of his knowledge on missiles from firsthand experience. Some of the missiles at the cape didn't scare me half as much as the small rockets Bill put together. He designed his own missiles using a steel pipe and used fuel from the real missiles at the cape! After launches, Bill searched through the fields for chunks of solid fuel which had fallen out during liftoff. With all the propellent he kept stored in his second-floor efficiency in downtown Cocoa, everyone knew if there ever was a fire in his room, the entire boarding house would be leveled.   As his expertise in missiles progressed, so did his designs. At one point, Bill even had the cape machine shop making parts for him.

There were numerous times when his missiles went out of control, such as the one he launched in my carport and another which crashed into a building at the cape. It amazed me that Bill and another cameraman, Les Case, never got caught shooting off their missiles on actual launch pads at the cape, filming the test with the Air Force's high-speed cameras and film. The two of them eventually formed a rocket club and moved their operations to a safer test site.

Of all the people I knew at the cape, no one came closer to being killed as many times as Bill. At least three missiles were detonated directly over his head, and another smashed into his camera site just minutes after he was called back to central control. And to this day, I can picture Bill and his many confrontations with animals: the rattlesnake that curled up on the seat of his camera mount and the two panthers who once refused to leave his site. I wished I would have been there to see his reaction when a cardinal landed on his head while he was filming a launch.

Range Rat Jim Sawyer wrote, *"I was in the army and spent '59 to '61 on the range, mostly on GBI (Grand Bahama Islands) but also considerable time on San Salvador and on the cape. We manned the radar sites that monitored the ionosphere in conjunction with the International Geophysical Year (IGY) but also supported the shots by providing ionospheric information to Central Control before, during, and after each shot.*

"I met a quality control engineer who had been one of the three people who manned command-destruct buttons at the cape. A missile would be exploded if, and only if, it didn't explode on the gantry; all three people pressed their button. This was so some out-of-control missile didn't go back over land.

"He told me about the time rock and roll music started playing on all the countdown boxes (we called them bitch

boxes . . . we had one at our cape site). We got bored listening to the countdown chatter and hooked our radio receiver to the countdown box so we could listen to both. In a couple of hours, the countdown boxes in our complex were shut down at a central telephone pole by maintenance workers. It was our radio that was playing, not just in our box but also in all countdown boxes at the cape.

"They investigated all the sites in our complex to find out what happened. Of course, by that time we had the receiver unhooked from the countdown box and were innocently sitting there performing duties at a high rate of precision. Then they turned our countdown boxes back on.

"We also used to set the radar on automatic, then climb the phone pole in front of our site so we could watch the launch just over the hill. They would blow up on the pad about half the time. The bitch box would scream, 'All range personnel, take cover!' We'd be there on the pole where cover wasn't available. There were instances of debris hitting while we scrambled down the pole."

Range Rat Chris Hummel, Sr., wrote, *"Made many trips on the C-124s. I flew on one of the first C-130-A on my second tour. On one of the runs from Recife to Ascension, the pilot put me in his seat and let me fly it for about 10 minutes. It was a RUSH! That would never happen today. The trip was so long on the 124s that I wondered if they ever would land.*

"Having been downrange when Freedom 1 flew and Shepard took us into space, I never heard one political comment in the four years I was downrange.

"Those who are now known as 'Range Rats' (I am proud to be one) just did superior work as part of a magnificent team whose purpose was to accomplish something that we had not done before. It was a great group who could understand that the success of the 'whole' (mission) is a result of the contribution of its 'parts.' As a radioman I was one of the ears that heard many things. We had fun; we partied; but when it came to the mission, what I heard was all serious business."

Range Rat Stan Wisdom wrote, *"I remember drawing straws to see who had to get out of the Air Force van to unlock the front door in the morning mosquito swarm!"*

Range Rat Ray Scarpa wrote about his flight on a DC-6, "*I flew on it many times. What a piece of junk it was . . . always broke down. One time we were broke down with a landing gear problem; and the base carpenter, Henry Wolf, got off the plane and fixed the gear with a piece of rope. Not sure what he did, but we took off and landed at the next stop without incident.*

"*Another time I walked down the aisle while in flight and saw one of the propellers stop turning. Good thing I was holding onto the two seats on either side of the aisle because the plane dropped down about 30' or so. Everything not tied down was up on the ceiling for a moment, including passengers. Yep, the good ole days.*"

Whenever any of the Range Rats working at the cape got together, we always exchanged our stories about the Navaho or "never-goes" as we called them. They seemed to blow up or go off course every time they were launched. I believe there were 10 or 11 in a row that failed or exploded. The United States knew Russian submarines, anchored within 10 miles of the Florida coast, were carefully monitoring our tests; so there was speculation that our radio signals were being jammed by the Russians. I don't know if there was any truth to that as it seemed it didn't take much to interfere with the Navaho's failures. I watched the last Navaho go up and just about fell off the phone pole I was on when it exploded over the beach. Yes, even with the known dangers, one couldn't help himself: it was *still* exciting to watch a rocket launch, and all the Rats climbed the poles to get the best view.

When I tell my stories of the close calls and living in miserable conditions in the palmetto scrubs of the cape to my friends, family, and groups to whom I spoke, they would ask if I would ever go through something like that again. My answer: I would go back tomorrow to do it all over again in a heartbeat . . . and I suspect any Range Rat would tell you the same thing.

_____ Chapter 13

I received training and operated several telescopes as they became upgraded. One of my favorites was known as the Boston University (BU) Telescope. The BU Scope originally arrived by barge at the cape, and I asked one officer, "What are you going to do with this?"

Boston University (BU) Telescope

He looked at me with one of those typical officer looks and responded, "The question is what are *you* going to do with it?"

I soon found out! The BU Scope had a 240″ telescope which was mounted on a large, Navy gun mount with four large wheels so that it could be pulled to anywhere it was needed. It had been redesigned to study atmospheric problems such as heat shimmer with which we had been encountering problems when recording the rocket launches on film. I remember during the countdown when we looked through our tracking scopes, the "bird" did a "shimmer dance" on the pad. The bigger the rockets, the higher they went and the further we had to get from the launch pad. So, the newer telescopes were a blessing.

B-36 Peacemaker

I didn't know much about the BU Scope when it arrived; but with all the training and practice tracking aircraft we did, I became pretty good with the darn thing. When rockets weren't being launched, we set up and filmed all the aircraft taking off and landing at the Air Force base.

At one point, the command staff decided to bring in a B-36, 6-engine bomber interestingly called the "Peacemaker" with target boards painted on the bottom of the wings. They flew over during different times of day and night in various weather conditions, temperatures, and altitudes; and we would record the target board on different motion picture film . . . 1,000′ rolls! This aircraft was massive, and it was hard to keep focused on the job of filming because it was so impressive to just watch, particularly during its flybys, coming in low over the ocean and across the beach then gaining altitude just to conduct another flyby.

My training continued when they built the new IGOR telescope. It was on top of a building with the beach as a backdrop. The IGOR was a much larger tracking telescope, capable of tracking missiles at a distance of up to 100 miles and providing photographic records of the missile's performance. Life seemed pretty good, sitting on the beach, watching and tracking rocket launches.

We worked closely with Air Force Intelligence during our assignments of tracking rockets and aircraft. On October 4, 1957, concern on the strained faces of the "big guys" in the inner circle and their rushed demeanors as they scurried around the base indicated to us something bigger than usual was going on. We learned Russia had successfully launched their first satellite, Sputnik 1, into space. At first no one was saying too much, but this had to be a major event. Russia was putting into space a satellite—something America had been unable to do. Russia's Sputnik 1 delivered a blast of shock and awe into the very core of our perceived defenses against Russia.

The perception that it wasn't a big deal was interrupted with a need for action. President Eisenhower signed the National Aeronautics and Space Act on July 29, 1958, and the development of NASA was underway.

Using IGOR, Ed and a technician prepare to track a launch.

The pressure continued as Russia launched Sputnik 2 on November 3, 1957, and Sputnik 3 on May 15, 1958. Although Sputnik 2 was only operational for around seven days, it continued to orbit until April 14, 1958. Sputnik 3 continued its orbit until April 6, 1960.

Of course, the United States rushed to enter the space race; we launched our first satellite, Explorer I, on January 31, 1958. What a wild time to be alive: witnessing and taking part in the space race.

My job continually rotated from filming and tracking the *local* rocket launches— which was exciting enough—to now having satellites cruising around in earth's orbit and their wanting us to track them.

Military intelligence determined when and where we went, what we did, and the assignments or missions assigned to us. I always had full confidence in these guys and loved working with them. I never really wondered whether they knew what they were doing because they portrayed such confidence and clarity that it gave us certainty in what they said. I had the utmost respect for the intelligence officers and all the military personnel with whom I was privileged to work.

As the Sputniks were being launched, one of the officers came to our site and said they wanted 35mm cameras with a split beam mounted on the radar dish in an attempt to get visual footage of the Sputniks. At first, I thought maybe it was a joke; but then I realized most military intelligence guys don't express much humor in their discussions. The next thing I knew, I was climbing the access ladder on the radar dish to see if we could cut holes in it. As with most other things we did, I rapidly realized this wouldn't be just an easy task: it was about 25′ in the air, and I almost slid off the s.o.b. because of the grease and oil that leaked off the radar. Fortunately, when I slipped, I caught hold of the access ladder. Even though I cussed them and said things I shouldn't have from time to time, I loved every minute of it. When all was said and done, we got two 35mm cameras mounted and working on the radar because as Range Rats that's what we did: we turned crazy ideas into reality.

When I went to work one morning, the Boston University Telescope (BU Scope) was gone, and I was told to go to the tech lab immediately. When I arrived, the two engineers with whom I worked were already there. There were also two majors from Air Force Intelligence, a pilot of a C-124 aircraft, and a sergeant from his crew. We were told that the BU Scope was the best thing that they had to try to get the configuration of Sputnik.

Soon came a copy of the military's special orders for several of us to travel. Around September 8, 1958, engineers Pete Button and Sedgwick King were joined by two intelligence officers from Wright Patterson Air Force Base and me as Chief Photo Electronics Technician to fly a mission with the designated purpose of operating the BU Scope, primarily to track the Russian Sputnik but also the first American satellite, Explorer 1. Although not known yet, this was the beginning of a project known as Operation Space Track.

We had a full military crew who packed up the telescope, electronics, and other equipment on a C-124 . . . but not without issues. The telescope wouldn't fit into the C-124. The scope was mounted on an old Navy gun mount, redesigned to

Military crew attempting to load a C-124 with the BU Telescope in preparation to travel to Pikes Peak.

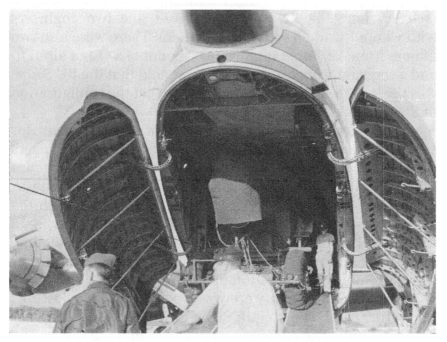

BU Telescope with deflated tires finally loaded.

be mobile for tracking rockets. Thank goodness for military ingenuity when one young officer suggested, "Why don't we just let the air out of the tires?" Sure enough, after following his suggestion, it barely fit.

We traveled from Patrick Air Force Base to Trinidad, St. Lucia, Antigua, Grand Turk, Mexico City, Elgin Air Force Base, Sacramento Peak, Reno, and Indian Springs Air Force Base, ending up on the top of Pikes Peak. What an incredible trip . . . first, heading down the eastern testing range, which we just called downrange, then across the country, ending up with an unexpected and unbelievable trip to the top of Pikes Peak with the BU Scope in tow. Can you believe they actually *paid* me for all this excitement?

The C-124 crew unloaded our supplies and equipment, preparing for our trek up Pikes Peak. Pete, Sedgwick, and I only had enough time to put together some general provisions and clothes for the trip. Headed downrange through the Caribbean, we were always ready with our Caribbean attire. However, upon learning there was snow and below freezing temperatures on Pikes Peak, we knew we were in trouble with our casual Florida attire. Thank goodness the Air Force always travels prepared as they soon issued us heavy pants and jackets with hoods to protect us from the snow and cold.

Soon they had the telescope loaded on a large military transport, and we headed up the mountain. We rode in the back of a military 4x4 enclosed by a heavily worn canvas top. The mood was light, and we were excited to participate in a new adventure: traveling to the top of a mountain to track satellites. One could feel the temperature getting colder with every switchback the convoy ascended; and before long, I spotted small ice patches on the side of the road. Having grown up in the Adirondacks, I was familiar with driving in snow and ice and wasn't overly concerned . . . well, concerned enough that I pulled the plastic covering back where I could see and speak to the driver and commented, "You know, this isn't Florida. These roads might get icy." The driver assured me he had it under control.

Looking out the back we could see down to the towns, which began to look smaller and smaller. The snow was building rapidly on the roadway. A short time later as we were talking about all the places where we had been, the truck turned to climb another switchback. It began to shudder a bit and started sliding backwards. There was nothing we could do but grab something to hold onto. Pete, Sedgwick, and I almost simultaneously let out a few expletives. I don't think we slid all that far; but at that moment, it seemed we were headed down the mountain over the side. Actually, we just ended up with one of the rear wheels in a ditch. As we bailed out of the back, the military personnel already were evaluating what needed to be done. One of the Air Force Intelligence officers told us to get into his station wagon. We traveled the rest of the way with him. I don't know if it was any safer, but it sure felt safer.

Arriving at the top of Pikes Peak, we found it covered in snow and cold. Our clothing provided by the Air Force made the cold conditions tolerable. It was a little eerie that we were the only ones on top of the mountain. A building stood a short distance away which I guessed was probably a small tourist center with a gift shop or small café. Of course, it was closed for the winter. There was another, smaller building in between

Unloading the BU Telescope at top of Pikes Peak

Work in progress on Pikes Peak

where we were parked and the tourist center which contained restroom facilities.

The Air Force Intelligence officers informed us of the time and coordinates that a satellite would travel above our position. It was time to get to work. Needing a place to store our film and to work on cameras and other equipment, the women's restroom was closest to us. So, the crew broke the lock on the women's restroom and figured out how to get a gas heater working, which was perfect for us to set up.

All our tracking and filming with the BU Telescope was done outdoors. We spent several days working in and exposed to the snow and cold, but no one seemed to mind. We Range Rats and the Air Force personnel all understood these missions were an integral part of making sure America would win the space race.

Many other Range Rats were assigned to find and track Sputnik 1. Wally Tubell, Sr., worked downrange in Mayaguez, Puerto Rico, and was later promoted to the Ship Instrumentation Manager, managing the ships in an out of the cape. He recalled the military and CIA being desperate to get some orbital tracking data on the Sputnik 1. Under clear-weather conditions, observers in the "right" locations on Earth could see sunlight reflecting off Sputnik's surface around dawn and dusk. This prompted the Eastern Test Range brass (and no doubt other agencies) to try to triangulate on those reflections. The brass quickly devised a plan for us to pair the Mod II radars to MK-51 gun directors, connect Giannini data recorders to the radars, and simply use spotters with their naked eyes and binoculars to look for Sputnik based on very rough "times" and "look angles." If the spotters were successful in their observations, it helped the MK-51 operator "acquire" the target.

Wally said he went out on almost every schedule with the guys to see if he could help spot the Sputnik 1 in the pre-dawn or post-dusk conditions. Wally proclaimed there were lots of laughs when they would mistake a car or plane in the distance for the "target" but claimed they did manage to track the Sputnik on a number of occasions.

One day they notified us that Sputnik was expected to pass directly over Mayaguez at an altitude of about 100 miles. So, the Mod II radar gang decided they would try to get some skin track on it (including range). None of the experts thought this would be possible. Nonetheless, the guys knocked themselves out fine-tuning the system. Luckily the MK-51 acquired the target, and the radar skin-tracked the Sputnik for about 90 seconds . . . another great accomplishment by the Range Rats.

Later in life I was fortunate enough to track down Harlan Spence, professor and department chair at Boston University. He responded to my email and said, "I do, indeed, have some history on that telescope/camera. My knowledge comes from two sources: an historical document of the Boston University Department put together by Professor Alan Marscher and from my dad, who was a student here in the pre-Sputnik era and worked with Professor Brigham (who retired just at the time that the camera/telescope you mention was being put into service)."

Mr. Spence went on to tell me that the key players regarding the BU Scope were "Lewis Brigham, Gerald Hawkins, and Duncan MacDonald. All were professors here at BU, the first two in astronomy and the latter in physics. The telescope was, indeed, built from an old Navy gun mount. At the time, my dad was a student here (in the School of Education). He interned with Professor Brigham. My dad was the assistant curator of the observatory. As such he worked some with Professor MacDonald, who was involved with the Optical Research Laboratory."

Ed with military intelligence and engineers and the BU Telescope

 Chapter 14

In 1957 Americans had just lived through the fear of Russia launching the first satellite orbiting over our heads. The anxiety was beginning to dissipate as the U.S. quickly countered Russia's satellites with our own successful program in 1958. Peace and tranquility did not last long when in October 1962 an American U-2 spy plane flying out of Patrick Air Force Base obtained surveillance photographs of Soviet, medium-range, ballistic missiles being built and installed in Cuba, a short boat ride from Florida. If one thought Sputnik 1, a small, basketball-sized, Russian satellite caused havoc, it would pale in comparison to what was getting ready to happen.

These missiles in Cuba were not much more than 90 to 100 miles from Cape Canaveral, our homes and families. If these nuclear-armed missiles became operational, they could easily strike targets in the eastern portions of the U.S. within minutes. President Kennedy decided to utilize the Navy to form a blockade around Cuba and delivered an ultimatum that the missiles be removed.

All the Air Force bases on the east coast of Florida were strengthened with B-52 bombers, fighter jets, military equipment, and personnel, arriving at a rapid pace. Aircraft were constantly landing and taking off from Patrick Air Force Base, and a steady presence of aircraft patrolled the beaches. Several hundred miles to the south in the Florida Keys they were lining the beach with barbed wire and establishing a fortress of radar, surface to air missiles, manned foxholes with machine-guns, anti-aircraft rockets, and other military weapons.

Many Americans felt that nuclear war may be imminent. Instances of amassing food, gas, and personal weapons was not uncommon. Yes, even I raced to J. M. Fields and purchased a gun. Standing in line with all the other frightened people waiting to pay, I thought, "What the hell do I think I am going to do with this gun? I live just miles from Patrick Air Force Base, now overstaffed with military personnel and weapons. Surely, if they can't stop an invasion of our beaches, did I really think I could do something with my new weapon?" But I still paid my money, leaving the store with my new gun in hand.

During my years at the cape, I worked for many different sections and areas on and off the cape and downrange. I like to think that I always got the calls and assignments because I was pretty darn good at my job. I loved being involved in space exploration and that I worked in concert with the military officers accomplishing their unique requests. But I also know that I was very fortunate to work in the heart of the cape where I became friends and a trusted confidant with a great man named Charles Buckley who everyone called "Charley."

Credit NASA

Charles Buckley at the STS-1 crew walk-out in 1981.

Charley was on active duty during World War II and became a Massachusetts State Trooper upon being honorably discharged. He became employed with Atomic Energy Commission Security Force at Los Alamos, New Mexico, in 1947, eventually transferring to the AEC San Francisco

Operations Office. Here he was responsible for security for shipments of nuclear material to the South Pacific and observed several nuclear tests at Eniwetok and Bikini Atolls. In 1960, Charley became the first Chief of Security and Fire Operations at Kennedy Space Center. Charley was in charge of not just the security of the entire facility, but also the security of all the astronauts and mission support staff. In many old pictures and media, one will see Charley eating meals with the astronauts prior to launch or escorting the astronauts to the pad on launch days . . . a job he not only took seriously, but which also gave him a tremendous sense of pride and honor.

So, how was I, a well-hidden, hard-working, field Range Rat so fortunate to become so well acquainted and trusted by the head of security operations? He bought a house on Jason Court in Satellite Beach one house from where my home was and became my neighbor where we developed a great friendship. Even though I knew the importance of the position he held, Charley always treated me and others with such great respect, never making you feel you were not an equal. He had great respect for all of the personnel at the cape, no matter what job we did.

I soon learned the sound of that familiar loud knock on the door by Charley which many times meant he had a new mission he wanted me to know about or be part of, and it wasn't always space-related. One particular Saturday morning I heard his loud rap on the door, and there was no doubt who it was.

I opened the door, and Charley didn't even say "Good morning." He just said "Ed, I bought a race car, and I want you to be my pit crew chief."

I said, "You did *what?*"

Charley again blurted out, "Seriously, I bought a race car . . . we're going racing!"

The next thing I knew, I was studying racing instead of telescopes and cameras; and sure enough, it wasn't long before we were in Valkaria, Florida, which is in Brevard County, for our first race. We recruited several other friends to help as the pit crew, and I must admit it was a blast. Charley was actually pretty good; but that shouldn't have surprised anyone as he

Driver Charley Buckley
with Ed as pit crew

March 1966: Buckley
and crew win First Place.

May 1966: the crew wins Third Place.

Another win for
Buckley and the crew

was dedicated to whatever he did . . . and, of course, we all got to go along for the ride with him.

Other fun assignments came with each knock on the door. One day Charley showed up and told me he needed me to pick up a couple of children of one of the Apollo astronauts that was currently on the moon. Charley gave me a piece of paper with the addresses from which I was to pick them up and to where I was to take them.

I asked Charley, "What am I supposed to do when I get them there?"

Of course, I got a typical, full-detailed, Charley description as he said, "You will see when you get there."

As a good soldier, I picked up the two children and took them onto the property at the cape and reported to the building given in the directions. I had seen this building on the cape property but had never been inside. As I approached the building with the two young'uns, a man came out and introduced himself as our escort. Inside the building we got into an elevator and rode the slow-moving elevator for what seemed like eternity.

As the doors opened, I couldn't believe my own eyes. It looked as though we had just arrived on the moon. With each step, it was as if we were walking on the surface of the moon . . . craters and all. I glanced up at the escort. He could see my astonishment.

He commented, "These kids will never forget this."

I looked directly at him and said, "Darn the kids! I will never forget this either!" . . . and I never did. It was an amazing feeling walking in there. It seemed so realistic. Although I was never told, I assumed it must have been a training simulation used by the astronauts.

Unfortunately, I do not recall which astronaut it was who was currently on the moon, but the escort told me he wanted his children to experience some of the feelings he was experiencing. The children were amazed. I know when we left, my eyes, as well as theirs, were still probably as big as some of the craters we had seen.

In 1970 the bottom fell out for me and thousands of others who were employed by RCA. The company lost its contract with the Air Force to Technicolor to provide launch data such as the pictures I was trained to take. Some of the people transferred into other departments, but most were laid off.

When we were all hired years before, we stood in long lines outside an old barracks, holding a urine sample in a plastic bottle, waiting to go inside for our physical. Now we were back in the same line, this time waiting for our last paycheck. What would we all do now? Funny, I never in the 13 or 14 years there thought of ever doing anything else for the rest of my life. I couldn't accept the fact that it was all over. Where could I go? A lot of people I knew there had to move away because there weren't any other jobs of any kind in the area. Some left in a desperate attempt to escape the reality of losing their job, their career. I was one of the few who decided to stick around. It took me several years . . . and a near breakdown . . . to recover from the shock. Even today I wake up at night thinking we're still in our house in Satellite Beach. But, as difficult as the transition was for me, I was one of the more fortunate people. At least I was still alive and fairly healthy. Some never got over it as they began drinking in order to dull their memory, and several of my close friends died.

To me that was the end of my days at the cape. I knew I had to forget about the past; otherwise, it would kill me, too. It was sad to see what the layoff had done to some of the other guys and also to the area. Stores and motels closed up. Houses were abandoned because people couldn't make the payments. There were no more traffic jams . . . even at rush hour. Things were like they were back in 1956 when I first came here. The only difference was me. I had no reason to be here now as I did then.

At first, I couldn't bring myself to open a box of newspaper clippings I saved about certain launches. Then I rationalized I had to write about it. Maybe I was bitter. I needed some way to get it out of me. I took on a variety of jobs. I still had a passion for photography and began covering NASCAR events, horse shows, and other events for several magazines. I opened a fishing bait shop and bought a large boat which I chartered.

Even after I left the cape, Charley's knocks continued as he reached out to me to escort the families of some of the astronauts to various events and photograph them and their children during the outings. After the horrific death of Astronaut Gus Grissom, I was privileged to be with Ms. Grissom at a racing event to which she was invited as a special guest and acknowledged. In 1971 Astronaut Cernan and his family traveled to Disney World where I was again thrilled to take photographs of their outing.

One afternoon, Jan was cooking when the familiar knocks hammered the door. As I opened the door, it was no surprise to see it was Charley.

"I have to talk to you *right now!*" he quipped. "It is important . . . very important."

I said, "Charley, can you give me about an hour? I will walk down to your house."

Charley all but growled, "Get back in the house! I need to talk to you *now!*"

Charley followed me inside and said, "Ed, I have to get back to the cape, and I need you to be there. We have a problem. There are terrorist threats against the Apollo 17 launch, the crew, and their families!" (See Link 1 at the end of this story.)

I reminded Charley that I no longer officially worked for the cape.

In his typical, in-charge style, he said, "I don't give a damn! I need you there!"

Charley was a great friend and mentor; and he knew if he ever knocked on my door, I would say yes which is exactly what I did.

He told me when and where to report. When I arrived, the guards already had my name and issued me a pass, providing instructions on where I needed to go. I arrived; and as I got out of the car, I could see Charley standing there with what

I called a "tommy gun" as well as other security personnel carrying the same type of weapons. If I didn't believe Charley before, I sure did now. I had never seen that type of security inside the cape before. As I walked toward the group, Charley came to meet me, repeating that there was a problem with terrorist threats.

"Ed, I would like you to spend time with the astronauts' families and photograph them while the boys head to space."

He introduced me and assigned me to work with cape security and members of the FBI in a group that never left the families. For the next eight to ten days, I attended all the events scheduled for the families, assisting them and photographing their adventures.

The night of the launch, Charley came to where the families were gathered and said, "Let's pray we get this done."

I remarked, "Let's just get Gene (Astronaut Cernen) to the moon where he belongs."

Charley walked me down closer to the group of people, some of whom were standing and some sitting in chairs.

He pointed and said, "That is Gene's family and others over there. I just need you to make sure the families are well taken care of and have everything they need."

I, of course, assured him I would; and Charley was off to his next task. Several members of the families eventually wandered over, and we engaged in general conversation about the excitement of their family member heading to the moon. I could never tell if the families knew of the threats or not because it was never mentioned.

It wasn't long before Astronaut Cernan, slated to command Apollo 17, showed up and spent a few minutes with his family. I knew Gene from my years on the cape; and as he walked by, he expressed his appreciation for my being there.

He said, "Take care of 'em, Ed."

I replied, "I will."

Gene said "I got to go; I got to go!"

With that, he was off with a security detail in tow.

In the group with the family that night at launch time was comedian Don Rickles who was friends with Astronaut Cernan and his family.

Soon the comedian came to where I was standing in the background and said, "I got to get down there closer. What are we doing so far from the rocket?"

I told Rickles, "That is the launch pad down there. You're not allowed to go down there."

Rickles grabbed his chair and said, "I got to get down there because I have to talk Gene through this," as he laughed and headed off toward the launch pad. He only walked a short distance toward the launch pad when in a loud voice so he could be heard by everyone, he started his routine as if he was talking to the Astronaut.

"Now look," he began boisterously, "I know you told me you were damn scared about heading to the moon, but I told you things will be alright."

He continued his hilarious act, typical of Don Rickles; and of course, the group was filled with laughter . . . including me.

At 12:33 a.m. on December 7, 1972, the 363-foot Apollo 17 lifted off safely under the command of Gene Cernan who landed and walked on the moon with Harrison Schmitt. The

Signed photo from Astronaut Cernan to Ed

two astronauts also spent considerable time traveling around on the lunar roving vehicle, returning to earth on December 19.

In January 1973 my wife and I received a wonderful, handwritten note mailed to our home from Jan Evans, wife of the Apollo 17 pilot, Astronaut Ronald Evans, thanking me for spending so much time with them and how she was looking forward to seeing all the pictures. Ms. Evans also thanked and acknowledged my wife's graciousness for tolerating my being away from my family. I always found the wives and families to be just as much of a class act as the astronauts themselves. Even to this day I have saved and cherished her thank-you note!

Signed thank-you note from Astronaut Evans' wife, Jan

_____ Chapter 15

It is sad that almost 40 years later America has not sent astronauts to the moon to continue the research and exploration. I hope with the renewed interest in space travel we will soon arrive there, as well as at many new, unexplored destinations.

I was privileged to observe and to work during historic rocket launches such as the Redstone; Navaho; Vanguard; Polaris; Thor; Snark; Atlas; Explorer 1; Saturn; Titan; Alan Shepard's first manned space flight aboard Freedom 7; John Glenn's first orbit in Friendship 7; the Mercury and Gemini flights; the Apollo 1 tragedy; Apollo 11, 12, 14, 15, 16, and 17 moon landings; and so many others in between. These brave astronauts climbed aboard rockets which launched them into the unknown to accomplish things that had never been achieved. They also had little assurance that their missions would be successful, returning them safely from where they

Ed, along with Sue Best, and John Glenn

started. As is well known, some of these heroes made the ultimate sacrifice, all in the pursuit of space exploration.

There are so many stories I would love to tell and to discuss involving these courageous men and women. The launches are well documented, and I encourage everyone to read and teach about all of them while I focus on the untold story of the workers.

Behind the scenes, the early space workers—from the scientists, engineers, technicians, and administrative personnel to the cleaning crews—provided the necessary support to make the missions happen. Some of these workers also made the ultimate sacrifice when catastrophic events or

Ed at space memorials with family

(L to r) Great-granddaughter Raegan, grandson Danial Hepker,
daughter Jennifer Hepker, Edward Ehrenspeck,
granddaughter-in-law Sy Lawing, and grandson Brandon Hepker.

Ed at memorial with great-granddaughter Raegan

accidents occurred in their work areas. All of these pioneers, their work and sacrifices, should never be forgotten.

Alan Shepard on May 13, 1996, said, "We need to remember the people who made it possible; so little is said of them."

Even today, people are fascinated, coming from afar and out of their homes to witness a rocket launch just as we all did 70 years ago. The next time you witness the miraculous launches and achievements that are being done today by both NASA and private companies, think about how it all came to be by a group of dedicated individuals who loved their country and did whatever it took to make America's space exploration possible.

I was thrilled when the city of Titusville and the U.S. Space Walk of Fame Foundation appropriated a site dedicated in 1994 and built the extraordinary American Space Museum and Space View Park which includes the U.S. Space Walk of Fame Memorials. (See Link 2 at the end of this story.) The memorials, which sit in the 2.63-acre park, are suitably located on the shore of the Indian River directly in line and just miles from the launch pad at the Kennedy Space Center and commemorate and highlight all the American space workers who have made today's space exploration possible. I am truly privileged to have my name inscribed on the Mercury, Gemini, and Apollo monuments.

Of extraordinary importance is the In the Line of Duty monument which memorializes the workers and astronauts who lost their lives in the pursuit and advancement of America's space exploration. We all should be indebted and never forget the brave individuals who sacrificed so much for our country.

On July 21, 2003, I attended and participated in the official ceremony and sealing of a time capsule which took place at

Ed teaching elementary children about space

Ed piquing the children's interest for space exploration

the Space Walk of Fame Museum. Astronauts and workers of Project Mercury placed their written memories and some memorabilia inside the capsule which resembled a Mercury space capsule and will be maintained at the Curriculum Materials Center Library of the University of Central Florida in Orlando. It is scheduled to be opened and the contents put on display in 2038.

My interest and love of rockets and space never diminished even though I had been laid off at the cape. In 1994 My two grandsons attended Killarney Elementary School; at the time I was retired. My daughter was very involved in volunteering at Killarney. My desire of sharing my love of space with young people led me to volunteer at Killarney Elementary, as well. Due to rheumatic fever when I was a young boy, I fell a little behind in school and, therefore, found myself as a father and a grandfather drawn to a particular group of students. At the time, they were labelled EH, emotionally handicapped. My daughter explained how their classes were separate from the other classes and that these students were prone to emotional and, at times, even violent outbursts. Well, that solidified it for me. I was going to work with these students.

I met Behavioral Specialist Kayla Garber, and I explained to her what drew me to and why I wanted to volunteer with these children. Although times have changed and inclusion is more evident in classrooms today, at that time these students were not. They were often shunned by the "regular" students and kept separate during some school activities. I remember thinking about these children and asking myself what could I do to help create a positive experience with and for them? The answer: build a rocket with them. Bingo!

So began a period of time where I truly believe I learned as much from them as they learned from me. Through Ms. Garber I was placed in one of the EH classes with Ms. Jennifer Roberts and Ms. Ellen Parker, two incredible educators. My first day volunteering in their classroom was quite an eyeopener. I

The rocket building begins.

walked into a room of about eight young men, one teacher, and one paraprofessional.

The first thing I noticed was what I referred to as "the boy in the box." No, he was not literally "in a box"! He was at a desk but had a three-sided barrier on his desk so that he was unable to view me or the rest of the students. It was explained quietly to me that he was trying to disrupt other students; therefore, this was in place until he could rejoin the class.

I began my introduction by trying to establish the fact that if we as a class were going to build a rocket and eventually launch this rocket that they needed to listen and follow the directions by me and their teachers. As an example, I played an audio tape of astronauts arguing with ground control and refusing to follow directions. After playing it, I explained that

I was in charge of the rocket building and launch and that their teachers were second in charge; but there would be no refusing to follow directions, or the project would end right then and there.

I also asked general questions about what the students already knew about space. Let's return to the boy in the box: for every question I asked, his little hand popped up. His knowledge was quite impressive so I knew he would be a help . . . and perhaps a hinderance. However, I was all in; and the rocket-building project began.

As with all children, they expected this to be completed in one day. It was necessary that I establish that we needed to learn about not only the rocket we were building but also about the tools we would be using and even the history of space flight . . . a lot of things would be covered; and it would, in fact, take time.

There were a few incidents. I sat with a couple of the boys while they painted the individual pieces of the rocket. All of a sudden, a chair flew past me. I froze to evaluate the situation. The boys painting didn't even flinch. Ms. Roberts took control of the situation with the disgruntled student.

One of the boys calmly looked at me and said, "He does that sometimes." With that, he continued painting.

Occurrences such as this did not happen often; but when they did, they were quickly handled and class continued.

The students were very excited as the rocket took shape and they learned that launch time would be very soon. They were taught about the fuel that would be used to launch the rocket, and they learned they would all have jobs. I assigned each student a job and trained him as to the tasks that particular job entailed.

But where to launch? The baseball/PE field was the perfect place, but there were PE classes out there all day.

As the NASA saying goes, "Failure is not an option." I met with the P.E. coach and explained what we wanted to do. He was 100% on board. All the right people were contacted, and all the rules were in place to be followed.

Launch Day! Word of the launch got around. Suddenly there were entire classes coming out to the P.E. field to watch the launch. Administration and teachers arrived for a good view, as well.

All the students in my group wore badges with their names and their title, which reflected what their job was before, during, and after the launch. However, suddenly the boys all

The launch team conducts their final checks.

And we have lift off!

got very quiet. I gathered them up to find out what might be the reason.

My heart broke a little when they told me, "You see, we EH students are looked at as being "different" than the "regular" students. We are never the center of attention in a positive way. What if this doesn't work? What if we are laughed at . . . again?"

Now when I said earlier that I learned as much from them as they learned from me, here and now in this minute was a perfect example.

So, I asked them the questions, "Did you build the rocket?"

"Yes."

"Do you each know your job?"

"Yes."

"Might this fail the first time?"

"Yes."

"Might it be a success the first time?"

"Yes."

"Now, look around you. They are all here to see *you*! So, now let's just forget about them and get this 'bird' in the air!"

And that is just what the launch team did! Once, twice . . . over and over again. And each time they did their jobs! Each time the rocket went up, and the retrieval team brought it back. And their audience cheered. In one short afternoon these boys—*my* boys—were the heroes . . . the victors. We packed up our equipment, and I think I saw them all walk a little taller as we walked back to the classroom.

Later that same school year I was told that the EH teachers and I had been nominated for the Orange County Public School (OCPS) Team/Teacher Volunteer of the Year Award . . . and we won! I think in the end there were a lot of winners, and I was honored to have been part of it.

This wasn't the only time I volunteered with OCPS students. I became a NASA- trained volunteer who was able to go into classrooms and teach about NASA to all the curious minds of the next generation. Interestingly they nicknamed me "Rocket Man."

I attended Science Project Night and set up rockets and handed out literature and talked to all the students who were interested in space and NASA. I spearheaded a field trip for Yvonne Holt's EH class of young boys and their para-professional: we headed to the Tico Air Museum and then to the NASA Space Camp. I addressed several classes at Killarney, ages kindergarten to fifth grade. I always brought all kinds of visual items for the students to see and touch, as well as stickers, etc., for them to keep.

Jennifer Roberts, Ellen Parker, and Ed
accepting Orange County Public School (OCPS) Team/Teacher
Volunteer of the Year Award

——————————————————————————— *Chapter 16*

I then moved to Edgewater, Florida, and contacted the Burns Science-Technology Charter School in Oak Hill which had just opened on August 22, 2011. There I met Dana Greatrex who made it possible for me to submit my proposal to the principal, Dr. Jan McGee, to set up a volunteer experience with their students. Principal McGee was busy trying to get the school up and running while I waited patiently for her to review the proposal. Actually, patience isn't one of my virtues. As I pestered her on a regular basis, she politely told me she had it on her list of things to accomplish. I think she finally moved me up on the priority list—not because of my program—to stop my relentless calls and visits. Finally, I was in!

My first visit, unfortunately, started out a little rocky. I was assigned an afterschool group, and I was just as excited to meet them as I had been all the other times I met new students, able to share with them my knowledge of and experiences with NASA. However, the students in my group proved to be quite a handful. The first meeting they decided to just run amok; and after a few minutes of that, I simply packed up my gear. I contacted the teacher in charge of them and told him I was leaving and heading home. Office Administration stopped me and asked me to return to the class. I declined.

I explained to them, "I take this part of my life very seriously. I want to make a positive impact on students. I'm sorry it isn't working out here."

To my surprise I was later contacted by Principal McGee and asked to come and see her. Upon arrival I was met with letters of apology from the students as well as confirmation from their parents that their previous behavior would not be repeated. I agreed to give it another try.

This time it was perfect, and I enjoyed a wonderful experience with these brilliant, young minds as we studied and learned about space, then built and launched rockets. Each of the youngsters carried out his jobs and mission assignments to perfection. Our afterschool sessions were terrific. In addition to the afterschool group, I expanded my volunteer time to teach day lessons on rocketry in their aviation learning lab and was soon contacted by one of the teachers about participating in the amateur radio club, as well.

So why not put them all together and see what happens? Through the efforts and encouragement of Mr. Josh Daniels with the Daytona Beach Amateur Radio Association and a lot of paperwork, the Burns School was one of 18 schools in the entire world selected to participate in the Amateur Radio and the International Space Station (ARISS) program. There was only six months to plan for the scheduled event even when

Volunteers build antenna at Burn Sci Technology Charter School

other similar events had one to two years of planning . . . and then not always successful. All that was known was NASA would provide a date in the middle of September based on orbital location and astronaut work schedules. There would only be one chance with no delays and no repeat efforts. A slew of amateur radio and other skilled volunteers began drafting a plan. As our date neared, two other schools had their hopes shattered as equipment failures on the International Space Station prevented contact. The volunteers built towers and antennas at the school and fine-tuned their radios. The students were allowed to write their own questions and then practiced asking the questions by using a HAM radio installed in the school with a HAM operator portraying an astronaut to communicate with them.

September 13, 2012, was the date assigned; and we had a huge crowd of parents, staff, and volunteers, including attendance by the Vice-Mayor and three city commissioners. At approximately 2:00 p.m., our time arrived.

"November Alpha One Sierra Sierra K K 4 K T J, can you copy us?"

The crowd listened with hopeful anticipation. The radio crackled with static, and finally Astronaut Sunita Williams responded, "We copy you . . . a little bit weak, but I can hear you. Over." Then Williams announced, "Welcome, Burns Sci Technology Charter School to the International Space Station."

My heart leaped, and I wanted to jump out of my chair. Everyone's efforts had culminated in this one, momentous, and unforgettable day.

I was like the proud parent of each one of those young students for all of their achievements that they took seriously and mastered. I was equally indebted to Josh Daniels and all his volunteers and Ms. McGee for taking the chance and making the time for a project that could have failed but was a major success.

The students were prepared to ask the astronauts questions.

First up, a young boy dressed in an astronaut outfit asked, *"How do you keep food cold up there? Do you have a refrigerator onboard? Over."*

Astronaut Sunita Williams

Williams replied, "We do have a refrigerator, but our food is usually at room temperature. Sometimes we prepare it with cold water and put it in the refrigerator, and sometimes we want something hot like soup and put hot water in it and eat it hot. Over."

Next up, student Tristan wanted to know, "*What do the astronauts miss the most about earth? Over.*"

Williams answered, "I really miss my dog, and today is his birthday. He is 11 years old. He is a little Jack Russell terrier; and when I take him for a walk, I usually feel wind on my face and sand on my feet at the beach . . . and those things about the planet I really miss. Over."

The next young student asked, "*What food do you miss most while in space? Over.*"

The astronaut responded, "We do not have a lot of bread up here because it floats around everywhere and even gets in our hair. So, I miss pizza, and I want to have pizza when I come home. Over."

Carli asked, *"What is the biggest city you can see from space, and how do you know it is that city? Over."*

Williams declared, "It is difficult at times to know which city is which, but sometimes it is easiest if they are on the coast or by the ocean or a lake. The brightest cities are over in the Middle East because people use a different type of lights. Over there it is very bright. Over."

Mia wanted to know, *"What is one project you are working on now? Over."*

Williams answered, "Wow, we are doing a lot of experiments on ourselves to find out what happens to bone density and our muscle mass up here, particularly in our legs, because we don't walk the whole time for months while we are up here. So, for me it is very interesting because it will help us to know what happens to people when they stay in space for a long time. Over."

Cali stepped up and asked, *"How do you get oxygen on the ISS? Over."*

Children talking to astronauts on International Space Station

Williams responded, "Great question! Some we get shipped up when we have some visiting spacecraft, but we also make oxygen by splitting water, (H_2O), into hydrogen (H) and oxygen (O) . . . and that is how we get our oxygen. Over."

Young Mickey said, "*Hello. Does being in space affect your metabolism, and are you hungry or less hungry in space? Over.*"

The astronaut replied, "The question was 'Does being in space affect your metabolism?' Yeah, we are trying to figure that out because people usually lose weight up here because you're not very hungry. Also, your body metabolizes food differently. For example, we don't need as much bone density up here so some of the calcium comes out of our bones while we're up here and we are trying to learn about that. Over."

Jarrod wanted to know, "*When you were on a spacewalk, were you ever afraid? Over.*"

Williams answered, "We have a lot to do when we are out there, so I wasn't really, *really* afraid. But at the end of the space walk, you want to really make sure you come in and don't forget your tools. We have to make sure we don't get tangled up because you are getting to the end of your limit of carbon dioxide and removal of oxygen, so that always makes me a little bit nervous. Over."

Tori asked, "*What do you have Robonaut2 doing on the space station now? Over.*"

Williams responded, "Robonaut2 has been doing all sorts of interesting things. Actually, he has what they call a 'busy board' where he can push buttons or activate switches. Two weeks ago, we had him wiping down a handrail with a tissue. Over."

The next young female student inquired, "*How do you take a bath in space? Over.*"

Astronaut Williams responded, "So very interesting question. Luckily it is not very dirty up here, but sometimes we get sweaty, though, because we work out or we run or bicycle or lift weights. You take a bath by putting warm, soapy water on a washcloth, then rinsing it off. Washing your hair is even a little more difficult than that because it takes some time, and you have to worry about the water going everywhere. Over."

A young, male second-grader wanted to know, "*What do I need to learn so I can become an astronaut? Over.*"

The astronaut replied, "I think some of the things are fun things like science and understanding some about math as those are some of the key elements of being an investigator like a scientist, veterinarian, or doctor. Those are the cornerstones, and from there you can do anything you want to do. Over."

Carli asked, "*What planets can you see from the ISS? Over.*"

Williams said, "Carli, that is a very interesting question. You know we are not that far; we are only about 250 miles or 400 kilometers above the earth; so relative to how far the stars are, we're not that much closer. So, we see the planets and stars much like you do. In the night sky you see Venus, Jupiter, or Mars. We see the same; they are just very clear. Over."

Mia wanted to know, "*What has surprised you?*"

The astronaut answered, "Wow, that is an interesting question. I think what has surprised me are the things that we do subconsciously. Like sleeping or eating sometimes is a little bit more difficult in space. I opened a bag of peanuts and lifted them above my head to make them fall in my mouth; and, of course, they floated out."

The astronaut's words were then interrupted by static.

Cali requested to know, "*Do you ever get claustrophobic; and if so, what can you do about it? Over.*"

Williams answered back, "I have never been claustrophobic. This spacecraft has about 10 modules, so it takes about 30 to 40 seconds to fly from one end to the other. So, it is pretty big. The spacecraft that brought us up here is a little bit smaller, but we only spend a couple of days in there when coming up to the space station and only about a half day in it when coming home. You are very busy; and when you're very busy, you don't feel like you get worried about that stuff. Over."

Mickey stepped up again and asked, "*How do you recycle water to drink on the ISS? Over.*"

Williams explained, "You use normal water that is shipped up here. You can even use pee—your urine—which we can clean through filters and make clean drinking water. Over."

Tori asked, *"Do you think mankind or families will ever be able to live in space? Over."*

Williams responded emphatically, "I think so, for sure. There are many things we still need to find out about . . . like what happens to the human body. But I think people will definitely be able to live in space and travel in space. Over."

Jarrod, stepping back up, asked, *"Who or what inspired you to become an astronaut? Over."*

Williams replied, "Wow . . . probably the spirit of adventure and exploration. I think all of us have it. I was in the Navy and the Navy tradition of astronauts which I heard about . . ."

Heavy static interrupted her answer, and the ISS continued on its mission. The call lasted a little over nine minutes with a total of 19 questions from those bright children. (Hear for yourself: see link 3 at the end of this story.)

The crowd in the room rose to their feet, cheering the students and all of the volunteers. Principal McGee took the microphone, describing the event as "awesome and makes me want to cry." She was so proud of her students and everyone involved. Principal McGee then praised and provided memory coins to each student and introduced and commended Josh Daniels and his HAM radio volunteers.

I also never forgot her generous words in calling me to the podium near the end of the event and proclaiming that what had just happened would never have occurred had I not brought the proposal to her, taught classes, and built rockets with the wonderful, young students and, hopefully, future space workers and astronauts.

Ed being recognized by Dr. McGee

Chapter 17

On February 15, 2013, the Volusia County Outstanding School Volunteer winners were named. To my surprise and honor, I was awarded the Volusia County Senior Volunteer of the Year.

NEWS 13 FLORIDA | VOLUSIA COUNTY PUBLISHED FEB. 16, 2013: "Volusia County Schools honors outstanding volunteers.

"Six outstanding school volunteers were honored Friday at the Volusia County School's annual VIPS Recognition Day. The event was held at New Smyrna Beach High School and honored volunteers who serve the county's public schools.

". . . Finally, Edward Ehrenspeck was honored as the outstanding school volunteer in the senior category. Known as 'Rocket Man,' Ehrenspeck became a trained NASA education volunteer upon retiring. Since then, he has provided numerous presentations and arranged for special tours and speakers for the classroom."

Several years later, Ms. Greatrex wrote, *"Because of Ed's promoting such a special connection with the school and NASA, an invitation was sent to invite Burns Sci-Tech students to attend another talking connection at the Kennedy Space Visitor Center with astronauts on a mission in space. Students and parents drove to the Visitor Center, spoke to several astronauts via KSC hook-up, and participated in a video that was shown on the NASA TV channel.*

"For several years, Ed formed an after-school Rocket Club. He supervised the students' building rockets and launching them on the athletic field. Today, the school has expanded to over 1,000 students, and a new high school building opens in August 2022. One of the career pathways is Aerospace, and partnerships continue to develop because of a personal door that Ed opened in a small charter school of 200 students."

From childhood on, I have always loved space, the stars, and the planets and was fortunate enough to work at NASA. It was very fulfilling to bring that experience and knowledge to the young girls and boys in both Orange County and Volusia County.

I am a member of the Elks Lodge 1557. In 2012 I was approached by two Elk members who are also members of the New Smyrna Museum of History. The museum has, in addition to its ongoing exhibit rooms, an additional room in which they switch out different exhibits. They approached me to see if I would like to set up an exhibit there about my time at Cape Canaveral. Would I ever!

I got to work, going through all my files and photographs. I contacted NASA in hopes of borrowing some items for the exhibit. I received a shuttle tire on loan from NASA along with some other items such as gloves worn by astronauts, handprints in cement, and a launch console from the Atlas Centaur. Items from my personal collection included photographs of the early rocket launches, original newspaper stories, patches, informational posters, personal photographs taken with astronauts, and rocket models to name a few.

The exhibit, "60 Years of History: 1950 to Present," opened, displaying the history of Cape Canaveral/Cape Kennedy. It was even more successful than I hoped.

The Daytona Beach News Journal printed the story, "The Canaveral Space Coast/ Kennedy Space Center Cape Canaveral Exhibit, which opened October 5, is a must-see. It is curated by retired NASA photographer, a current volunteer of NASA education

Museum
space exhibition
extension
announcement

Ed, curator at the museum

serving on the Board of Education and lifetime
New Smyrna Beach resident, Edward Ehrenspeck.
Ehrenspeck played a crucial role in NASA two years
before it was built in Titusville. He is available at the
museum every day to answer any questions and give
fascinating accounts of his experience at NASA and
of his longtime relationships with the astronauts."

Every day from 10:00 a.m. to 4:00 p.m. Tuesday through Saturday I arrived at the exhibit ready to greet the people and share my passion for space, rockets, and my experiences from the early days and through my years working at the cape. I had a blast!

Local people, men and women who also had spent time working at Cape Canaveral, came and talked about their experiences, swapping stories and memories with me. Young adults and teenagers, who only knew about the years around the space shuttle and were often surprised by the history as well as the sweat and blood put in long before the space shuttle, also came. People who were angry that the space shuttle had been discontinued were eager to hear why and how the shuttle was used mainly to transport everything up into space so the space station could be built and how the next phase of space travel and the collection of knowledge would begin and then be built upon.

Then there were the children, wide-eyed and excited when I gave them a coloring sheet, a paper shuttle to cut out and build, stickers, or pictures. I always made sure I gave the same attention to a curious child as I did to a curious adult. After all, the children were the next generation who needed to become interested enough to get involved with space exploration.

Challenges came when visitors from other countries visited and wanted to know about my exhibit and material.

Ed teaching at the museum

Not to worry though . . . I remember several times when hand gestures, photographs, and a little bit of attempting to communicate verbally got anyone who loved space past any language barriers we might have had.

Suddenly word got out, and people came to visit the exhibit and museum or came to visit the museum and stayed to see the exhibit. Next thing I knew, I was asked to extend my exhibit through November and December so people could bring their family members to the museum during the Thanksgiving and Christmas holidays. Amazingly it was again extended to June 15, 2013. This turned out to be the longest-running temporary display in the museum's history!

To say I was proud and excited would be putting it mildly. I was so honored that my exhibit regarding my time at Cape Canaveral and my love of rockets and space exploration was so well received. I met so many wonderful and interesting people during that eight-month period. Actually, I met so many wonderful and interesting people during my *entire career*! Wow, what a ride!

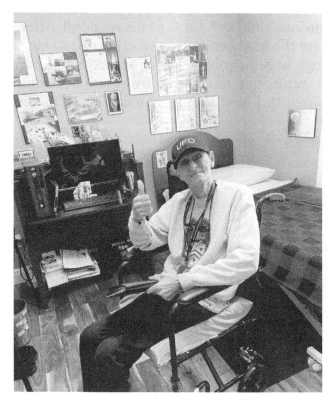

Ed still teaching from his room at age 92.

 A Note from the Family

One of Ed's goals in life was to complete and publish his book so everyone would know and never forget the sacrifices of all the early space workers and astronauts who made current, frequent space exploration possible.

Ed was hospitalized on March 8, 2022, and transported home on March 12, 2022. The hospital warned he may not survive the transport. When Ed arrived, he was having difficulty speaking and began pointing at his pictures and giving the ambulance attendants a lesson about early space travel. Remarkably they stayed about 20 minutes, listening and asking him questions.

Hospice Nurse Marla arrived several hours later, and Ed rallied again . . . another potential student. Now, sitting up and speaking in a strong voice, he instructed Marla to take down pictures so he could begin to educate her about space. For the next two to three hours, he narrated all his stories to her. It was amazing to see as Ed never stopped teaching about his love of space exploration.

Unfortunately, Ed began to decline soon after his last lesson. He did not get to see his book published as he peacefully passed away on March 16, 2022.

Interestingly, Nurse Judy arrived for her shift and was only with Ed for about five minutes when he passed. When things calmed, Judy looked over all his memorabilia on the walls.

She said, "When I lived in New Smyrna Beach, I went to the museum where a man was explaining all of this."

Jennifer asked her, "On Canal Street?"

The nurse said, "Yes!"

They both realized at that moment that it was Ed who had been the curator.

Ed loved, cherished, and spoke so highly of all the astronauts, military personnel and officers, space workers, and the Range Rats up and down the range. He held every one of you in such high regard; and, as he said in the book, he would do it all over again, given the opportunity.

A special thanks to the hospice nurses Marla, Nan, Shana, Laura, Lisa, and Judy who helped prepare Ed for his final flight and to Mr. JanNavis from the funeral home who, after seeing and hearing Ed's space history service and love of country, carried him draped in the American flag to his vehicle, which Ed would have loved and been in his glory to see.

The man, Edward H. Ehrenspeck, who loved aerospace and was affectionately known by students as "Rocket Man," will fittingly be making his final journey into space: working with Celestis, his remains and the DNA of Sue Best, his 26-year, loving companion, will travel forever together as they board the Excelsior Flight from Cape Canaveral on a Falcon 9 rocket launched by SpaceX projected to occur in the first quarter of 2023. Ed will orbit the earth on a satellite released from the Falcon 9 rocket and begin his travels and exploration of new destinations. We hope that his space travel will end in a place filled with love, happiness, and peace—a fantastic tribute and closing for a NASA Range Rat.

"Time to ride what you helped create."

From his family, we thank all of you.

Links:

1 Magazine Article, "Target America 1972: When terrorists Threatened Apollo – An Untold Story of Apollo 17 by David Schlom:
https://www.google.com/url?sa=t&rct=j&q=&esrc=s&source=web&cd=&cad=rja&uact=8&ved=2ahUKEwj6ota6iLj4AhWtGVkFHTjSBT8QFnoECAQQAQ&url=https%3A%2F%2Fspace.nss.org%2Fwp-content%2Fuploads%2FAd-Astra-Magazine-v13n6f5.pdf&usg=AOvVaw0I3U2DCvt9z0TTu4vAVf6G

2 American Space Museum and Space View Park:
https://spacewalkoffame.org/

3 Burns Sci-Tech Charter School Talks to the ISS: https://www.youtube.com/watch?v=RkrTHAE8zBw

Made in United States
Orlando, FL
08 August 2023